THE SACHERTORTE
ALGORITHM

THE
SACHERTORTE ALGORITHM

AND OTHER ANTIDOTES TO COMPUTER ANXIETY

JOHN SHORE

VIKING

VIKING

Viking Penguin Inc., 40 West 23rd Street,
New York, New York 10010, U.S.A.
Penguin Books Ltd, Harmondsworth,
Middlesex, England
Penguin Books Australia Ltd, Ringwood,
Victoria, Australia
Penguin Books Canada Limited, 2801 John Street,
Markham, Ontario, Canada L3R 1B4
Penguin Books (N.Z.) Ltd, 182–190 Wairau Road,
Auckland 10, New Zealand

First published in 1985 by Viking Penguin Inc.
Published simultaneously in Canada

The assertions contained herein are those of the author and are not to be construed as official or reflecting the view of the Department of the Navy.

LIBRARY OF CONGRESS CATALOGING IN PUBLICATION DATA
Shore, John.
The Sachertorte algorithm and other antidotes to
computer anxiety.
1. Computer literacy.   2. Computers—Psychological
aspects.   I. Title.
QA76.9.C64S56   1985   001.64   84-48843
ISBN 0-670-80541-6

Acknowledgment is made to the following for permission to reprint copyrighted material:

Association for Computing Machinery, Inc.: Portions of a letter from Peter Buhr that appeared in the ACM Forum section of *Communications of the ACM*, 26:7, July 1983, pp. 463–64. Copyright © 1983 by Association for Computing Machinery, Inc. Selections from Edsger Dijkstra's "The Humble Programmer," published in *Communications of the ACM*, 15:10, October 1972, pp. 859–66. Copyright © 1972 by Association for Computing Machinery, Inc. Selections from Donald Knuth's "Computer Programming as an Art," published in *Communications of the ACM*, 17:12, December 1974, pp. 667–73. Copyright © 1974 by Association for Computing Machinery, Inc.

Richard P. Feynman: "There's Plenty of Room at the Bottom," transcript of a talk given by Dr. Feynman on December 29, 1959, at the annual meeting of the American Physical Society.

McGraw-Hill, Inc.: Selections from *Elements of Programming Style*, Second Edition, by Brian W. Kernighan and P. J. Plauger, McGraw-Hill, Inc., 1978.

Macmillan Publishing Company: A selection, adapted from the table of contents, from *The Elements of Style*, Third Edition, by William Strunk, Jr., and E. B. White. Copyright © 1979 by Macmillan Publishing Company.

Perfect Software, Inc.: From the *Perfect Calc Manual.* Used by permission of Perfect Software, Inc., Berkeley, CA 94710.

Springer-Verlag, Inc.: A selection from *Selected Writings on Computing: A Personal Perspective*, by Edsger Dijkstra, Springer-Verlag, Inc., 1982.

Printed in the United States of America
by The Book Press, Brattleboro, Vermont
Set in Century Expanded

To my parents:

Hanna Fischmann Shore
Felix Francis Shore

Acknowledgments

I wish I could say that this book had seethed within me for years, that I wrote it when I could no longer prevent myself from doing so, and that it sprang out, wholly formed, in a week or two of frenzied writing. In fact, it wasn't even my idea. Raphael Sagalyn suggested it. I thank him for doing so and for helping me throughout the project as my literary agent. Without Rafe, this book would not exist.

I thank my friends and my family for responding with grace and generosity to numerous impositions during the past year. I'm especially grateful to Susan Bachurski. With her tolerance, her support, and her help, I wrote a better book, I wrote it faster, and I enjoyed it more. I'm likewise grateful to my daughter, Hilary. She too was tolerant, supportive, and helpful. And, like Susan, she loved me throughout.

When Hilary first heard me speak of my deadline, she asked, "Is that when they kill you?" Fortunately, Martha Kinney chose a less literal interpretation; Martha edited the initial drafts of this book, and she understood the need for gestation. Using her considerable skills as a literary surgeon and literary nurse, Martha helped me to get the book off the table. But it took even more to get it out the door. I'm grateful to Gerald Howard, who edited the book during its final stages and saw it through production; and I thank the many people at Viking Penguin who helped to shape the book's form and refine its contents.

Apart from my editors, a number of people read all or part of the manuscript and took the time to give me comments. I thank them all for doing so, especially Joan Bachenko, Susan Bachurski, Paul Bennett, Dick Blume, Rod Johnson, David Lohman, Dave

Parnas, Jane Quinn, Terry Quinn, and Rafe Sagalyn. Thanks also to Carla Allen, Pat Bronstein, Don Foley, David Johnson, Grace Reef, Janet Stroup, and all of the other people who helped me in various ways.

That I knew enough to write this book at all is the result of a stimulating, varied, and supportive research environment at the Naval Research Laboratory. For these advantages I thank the laboratory's management and staff, as well as my colleagues over the years. During the past year many colleagues contributed directly to this book by answering questions and participating in discussions, and I thank them for doing so. I'm especially thankful to David L. Parnas. During our long association Dave stimulated my interest in software engineering, and he introduced me to many of the ideas that I present in this book. Moreover, many of the book's details were shaped by the numerous conversations we had while I was writing it.

The book is not a work of history. Accordingly, except when citing some key developments and except when quoting individuals, I give few specific credits. But much credit is deserved, and it's a pleasure to thank generally the many people who have discovered, explored, and illuminated the book's subjects. It's my viewpoint, but I describe what they accomplished.

Contents

Preface

There are, you may have noticed, other books about computers. Why another one? Authors are often warned not to take the considerable trouble of writing a book unless they have something to say that's new and needed. Having just finished, I can vouch for the trouble. As for something new and needed, let me first point out what this book is not. Namely, it's neither a market survey nor a "how to" manual; it won't tell you what to buy, and it won't tell you what to type. Rather, it seeks to promote a more general understanding of computers. Of course, other books have this same goal. But many of them are technically formidable, and those that are more accessible tend to be either superficial in their coverage or unnecessarily antiquated. This last complaint may seem odd in reference to such a modern technology, but we've learned a lot about computers in the last twenty years, and much of the information is both useful and easy to understand. Besides, it's interesting. Modern developments in computer programming have uncovered some of the most important, fascinating, and intellectually challenging problems of our time.

"User-friendly" is likely to go down as the advertising talisman of the 1980s, and its prevalence may make you wonder why anyone but the technologically curious still needs a book about computers, even this one. One answer comes from comparing computers with two other examples of modern consumer electronics: stereos and CB radios.

A childhood neighbor of mine talked a lot about his tweeters and his woofers. He wasn't talking about his birds and dogs; he was talking about his hi-fi system. My neighbor was a hi-fi hobbyist, which meant that he was a cross between a music lover and an electrical engineer. He talked about clarity and timbre, just as he talked about impedances and roll-offs. In those days, getting the best sound from a hi-fi often depended on knowing how it worked. This was an advantage—my neighbor got as much pleasure from designing, assembling, and adjusting his hi-fi as he got from listening to it. Ham radio operators are similar. Many of them have built their own transmitters and receivers, and much of their pleasure in doing so results from the exercise of technical knowledge. Indeed, the demonstration of technical knowledge is a legal requirement for an amateur radio license. For a hi-fi buff the "end product" is music, and for a ham radio operator it's communications. But in both cases—as in other hobbies—the process is as important as the product.

The hi-fi hobbyist is a dying breed. Today's sophisticated stereo systems are, to use that trendy term, user-friendly. Roughly translated, this means that they're easy to install and easy to use. To exploit their capabilities you don't need to know how they work— you only need to know how to spend money, plug in connectors, and push buttons. CB radios are likewise a user-friendly form of amateur radio.

To an extent, computers have evolved similarly. When personal computers first became available, computer hobbyists proliferated. They struggled with the likes of bits, bytes, boots, CPUs, instruction sets, and assemblers. They thrived on this struggle, and they thrived on its requirements for technical knowledge. Their "end product" was a computer system that worked, but, again, the process was as important as the product. Lately, as manufacturers seek to expand their markets, the user-friendly trend has set in. The goals are the same as with stereos: plug them in, turn them on, press a few buttons, and enjoy.

But these goals are elusive. Today's computers are much less user-friendly than consumers want and manufacturers claim. There are cracks in every user-friendly facade, and the cracks make it hard to use computers without understanding something about how they work. Few are satisfied with this situation, but it's a fact of today's technological life. The technical problems involved in eliminating the cracks are extremely difficult, and their solutions will arrive slowly.

Unfortunately, many people can't wait. The computer is arguably the most flexible, capable, pervasive, and important tool that has ever been invented; people want and need to deal with it. For some it's a matter of getting ahead; for others it's a matter of catching up. Whatever their reasons, many people are reluctant; they want to learn about computers, but they're not champing at the bits.

They may be afraid. Anxiety about computers is caused by more than their newness, their apparent complexity, and their threat to the status quo. There are important, more subtle causes, having to do with technological intimidation, with basic differences between people and computers, and with assumptions that people make about the correctness of computer outputs. Computer anxiety is widespread—people are feeling it, the media are reporting it, and psychologists are studying it. Many people are intimidated, frustrated, and annoyed by their computer experiences. Other people may not use or want to use computers themselves, but they're concerned about the implications of this life-changing technology; they want to know more about it.

All of these people need information. Information about computers can help to overcome anxiety, and it can give new users the ability to patch the cracks in user-friendly facades. Information can help more experienced users to improve their capabilities. And for those who are concerned about the effects of computers, information is what's needed for informed opinions.

I hope that this book will help. I've organized it into three parts. Part I has a single chapter, one that's more about our reactions to computers than it is about computers themselves. In particular, I consider the common tendency to be intimidated by computers and computer outputs, and I consider various forms of computer anxiety—anxieties about using computers as well as anxieties about their effects. I discuss some causes of computer anxiety, some means of coping with it, and the benefits of doing so.

Part II, Chapters 2 to 7, comprises the equivalent of a short course in computers. But the style isn't that of a textbook. Rather, as in the rest of the book, I present the information in a series of relatively self-contained essays. Part II begins with a general discussion of what's good and bad about computer jargon, and it ends with a gentle introduction to the subject of programming. In between, I explain the principal components of computer systems, and I explain why computers interact with us as they do. The extent to which these interactions seem to require unfamiliar technical knowledge is a major difficulty for many new users, especially those who approach the computer looking for a new tool rather than a new hobby. But here's a case where it's easy to acquire a little knowledge that can go a long way. Some insight about what goes on behind the screens makes it much easier to sit in front of one.

While Part II focuses on a general understanding of computers, especially from the point of view of personal use, the focus of Part III is much broader. It covers the reliability, the technology, and the ultimate capabilities of the huge computer programs on which our society depends. Much of Part III concerns a phenomenon that has become widely known as the "software crisis"—our inability to write large computer programs that we know to be correct. This is hardly an academic issue, since our financial systems, our energy systems, and our weapons systems all depend crucially on such large computer programs. In Chapters 8 to 10 I describe the software crisis, I explain its major causes, and I introduce you to some of the technology that may enable us to cope with it. Finally, in Chapter 11 I turn to the fascinating field of artificial intelligence— a field in which people are attempting to create computer systems that emulate human intellectual activity. I discuss the differences between human brains and electronic computers, and I explain why I believe the line between science fiction and fact to be far from the well-publicized position of the artificial intelligentsia.

js
Washington, D.C.
1984

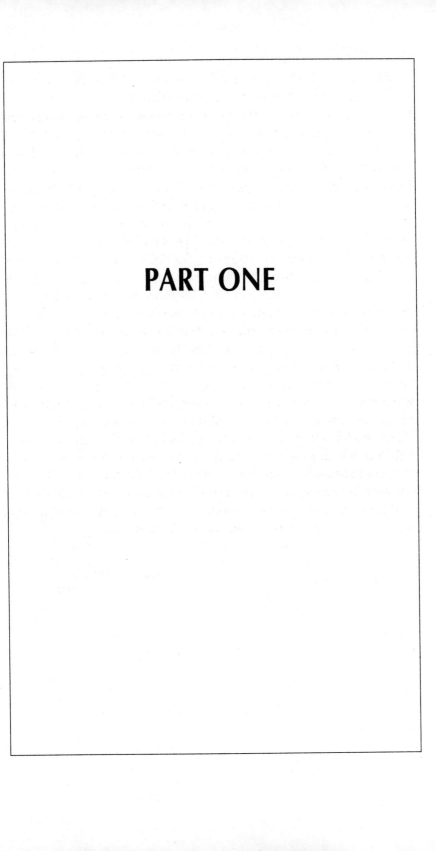

PART ONE

The first commercial computer was installed less than thirty-five years ago. The first of the so-called desktop or personal computers was sold in 1975. Today there are millions of computers in use, and millions more are sold every year. They are at work and at home, in our tools and in our toys. So much depends on them—energy supplies, weapons, institutions, offices, assembly lines, finances, and fun. Likewise transportation and communication—whether we move ourselves or information, we depend on computers. Not since the invention of the printing press has a new technology had such far-reaching implications. Never has a technology affected so much so fast. No wonder computers are intimidating. No wonder they cause anxiety.

CHAPTER 1

Intimidation and Anxiety

intimidate, *v.* to render timid, inspire with fear; to overawe, cow . . .

anxiety, *n.* uneasiness or trouble of mind about some uncertain event . . .

—Oxford English Dictionary

Like some other first experiences, my first computer experience occurred during my undergraduate years. Then, my lasting interest in computers began not as the result of carefully planned Ivy League diversity, but as the result of belonging to the Yale Flying Club. I was an "active member"—roughly translated, this meant I would rather fly than study. But I never flew as much as I wanted to, primarily because it was expensive and not included in room, board, and tuition. My parents supported my education generously, but they drew the line at flying, so most of the time I had to support the habit with my own earnings. This requirement led to a succession of odd jobs and finally to the offer of a part-time job as a research assistant in the physics department. The job was a real plum, especially since I was a physics major, so I gave up the mixed blessings of my previous position as a janitor at a nearby girls' school.

The new job was my first direct encounter with nontextbook science, and as such it had all the formative aspects of a novitiate. I was lucky in having a boss who made the experience a positive one. Dr. W. Raith was an atomic physicist with a special interest in experiments performed with beams of electrons. I helped make up the "targets" that were placed in the electron beams, plotted

4

graphs of data from experiments, fetched articles from the library, and performed a variety of other tasks befitting my position. At one point Dr. Raith asked me whether I could learn enough computer programming to calculate the predictions of some theoretical models. Of course I said yes.

I was delighted to have someone pay me to learn about computer programming. I was also excited that Dr. Raith and others might rely on my programs. This would be analogous to designing and building equipment for their experiments, and for an undergraduate assistant to do this, rather than merely to operate the equipment, was unusual. I wish I could say that I was so honored because of my brilliance. In fact, other, more appropriate people were too busy, and Dr. Raith didn't know how to do it himself; moreover, he was uncomfortable about learning how. His reluctance was my first encounter with computer anxiety.

Computers and Man at Yale

In the mid-1960s, the center of computing life at Yale was the Thomas J. Watson, Jr. Computer Center, an appropriately modern-looking building that housed several large IBM computers. The computers were accessible to all, but only visually. They sat behind a long wall of glass, where we watched them receive the attentions of full-time attendants.

In those days few people interacted with computers directly by means of computer terminals. Most people prepared their programs and data by using card-punch machines to punch holes into a series of oblong cards. Punched cards were first used to control a machine by the Frenchman Joseph Marie Jacquard. In 1805 he invented a device that attached to a loom and automated the weaving of complicated patterns, and the Jacquard loom quickly became important to the textile industry. Punched cards were first used to aid computation by Herman Hollerith, who used them to simplify the tabulation of the 1890 U.S. Census. Later he founded a company that eventually became the IBM Corporation, and such punched cards became known almost universally as IBM cards. Today they've been largely supplanted by other media—IBM cards aren't used in word processors and personal computers—but they haven't disappeared entirely. For example, they're still commonly used in computer systems that issue and process payments in the form of checks. Anyone who has ever received an IRS refund or other government check has held an IBM card. The phrase "Do not fold, spindle, or mutilate" came from the IBM card and was at one time symbolic of the computer age.

Whenever I "punched up" a program at Yale's computer center, I would place the resulting deck of cards in a special tray, where it waited, squeezed between other people's card decks, until the computer exhausted its current backlog. This could take anywhere from a few minutes to a few hours. Often we would just hang around, watching through the glass.

There wasn't much to see. There *was* mechanical activity—cards disappearing into machines, cards appearing from machines, tapes spinning, and paper being printed—but all of this had to do with getting data in and out of the computer, and none of it had much to do with what was going on inside. The only visible evidence of actual computing was a shimmering array of lights, each one blinking on and off as various changes occurred within the computer. Most of us had no idea what the lights meant, but we stared at them anyway. Scenes like this were common at computer centers throughout the country. Rarely have so many stared so long at so little visible activity.

I would watch the card-reading machine, trying to tell when my deck was being read. At the critical time my eyes would shift to the shimmering lights, as I tried to imagine the computer working on my program, and then to the printer, where I could often spot the results being printed. It was like watching a magic show.

My understanding stopped at that glass wall. I understood, albeit incompletely, how my computer program posed numerical questions and defined a procedure for answering them. But when the answers came back, or when the program was rejected for one reason or another, I was always a bit surprised. I didn't have the faintest idea of how the computer did what it did. People told me that the underlying principles were simple, but I didn't believe them. I knew that, barring some malfunction, the resulting printout was strictly a function of the card deck I had submitted, but I couldn't shake my sense of mystery, even when the printout was what I expected. And when the computer rejected my program, it would accompany the rejection with a series of cryptic pronouncements that I was unable to decipher. Off I would go, seeking divination at a special desk provided by the computer center to deal with this common phenomenon. There, one more versed than I would interpret the computer's pronouncements.

The glass wall reflected our fascination and it catered to our enormous curiosity about this new technology, but it also emphasized our separation from it. We could look, but we couldn't touch. Looking longer achieved familiarity, but not insight. By emphasizing our separation and our lack of understanding, the glass wall reinforced our tendency to be intimidated.

The Barrier of Miniaturization

"It's like magic" is a common reaction to a demonstration of the computer's abilities. The urge to believe in magical and psychic powers is strong and well documented, and there's no reason to think that the urge doesn't extend to computers. Most of us, however, don't believe in magic, although we often enjoy watching magicians. We think of them as having technical skill rather than magical power, we speculate on the mechanism of deception, and we want to see how it's done. So it is with the computer.

When a machine's moving parts are visible, they help to make apparent the machine's logic. When you turn the steering wheel in your car, you rotate a shaft that moves some rods that turn the car wheels, all of which changes the direction of travel. But if you look inside a modern computer, the most activity you're likely to see is that of a fan pushing air, if that. Electronic logic replaces mechanical logic. There are no moving parts, only moving electrons. It's hard to develop intuition about moving electrons because their movements are invisible, and their effects are statistical.

Modern information processing is not only nonmechanical, it takes place on a microscopic scale. It wasn't that way at first. The first electronic computer, the ENIAC (Electronic Numerical Integrator and Computer), was housed in a room 30 by 50 feet, weighed 30 tons, and contained more than 18,000 vacuum tubes (Figure 1). The ENIAC was dedicated in 1946. Today, you can carry a much more powerful computer around under your arm, and you can balance its main internal units on your fingertip (Figure 2).

Miniaturization is a barrier to the senses and therefore a barrier to the acquisition of physical intuition. And without physical intuition, it's hard to feel comfortable about microscopic engineering. The microscopic scale of modern electronics has a lot to do with the computer's technological intimidation; it's hard to believe that so much can take place on such a small scale.

"There's Plenty of Room at the Bottom"

In fact, there's just as much room for microscopic engineering as there is for macroscopic engineering. The head of a pin, for example, may seem small to us, but there's enough room on it to write the entire *Encyclopaedia Britannica*. And when I say "write," I don't mean in terms of some computer-readable code, I mean directly, with letters and pictures.

The *Britannica* example is from an essay by the American physicist Richard Feynman. His subject was the problem of manipulating and controlling things on a small scale, and his message was summarized in his title: "There's Plenty of Room at the Bottom." The *Britannica* example helps bring the small scale of the microscopic world into focus. Here's Feynman's explanation:

The head of a pin is a sixteenth of an inch across. If you magnify it by 25,000 diameters, the head of the pin is then equal to the area of all the pages of the *Encyclopaedia Britannica*. Therefore, all it is necessary to do is to reduce in size all the writing in the *Encyclopaedia* by 25,000 times. Is that possible? The resolving power of the eye is about 1/120 of an inch—that is roughly the diameter of

Figure 1. The ENIAC computer at the Moore School of the University of Pennsylvania *(Photograph courtesy of the Smithsonian Institution)*

OPPOSITE: Figure 2. A typical modern computer chip. A chip like this can contain all of the arithmetical and logical functions of a simple computer, and a few other chips can contain all of the computer's memory. *(Photograph reprinted by permission of Intel Corporation, 1984)*

Figure 3. Food for thought *(Photograph courtesy of North American Phillips Corporation)*

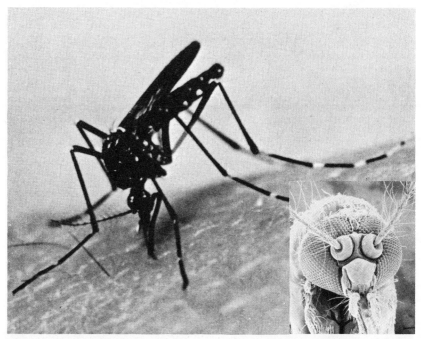

Figure 4. Although the mosquito would prefer the finger, it could sit easily on the chip in Figure 2. This airborne hypodermic needle comes complete with visual and olfactory sensors, together with information-processing and flight-control systems that direct the needle with uncomfortable effectiveness. Human achievements that compare with this occur on a much larger scale (see Figure 5). *(The full-view photograph is courtesy of Jack C. Jones. The inset electron micrograph is courtesy of Ralph E. Harbach; from R. Harbach and K. Knight,* Taxonomists Glossary of Mosquito Anatomy, *Plexus, 1980, reproduced by permission.)*

OPPOSITE: Figure 5. The Apollo Lunar Lander *(Photograph courtesy of NASA)*

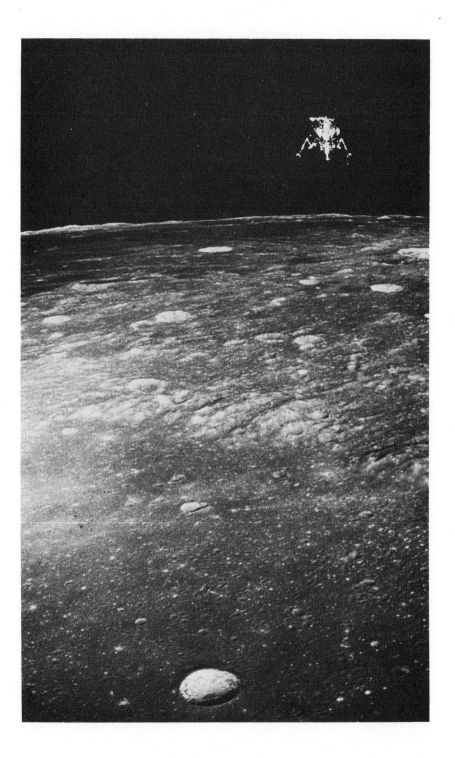

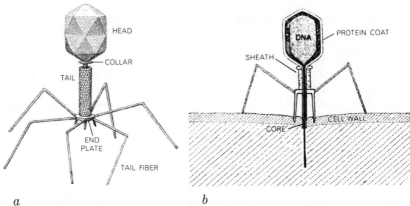

a   b

Figure 6. Diagram of a T4 virus. The T4 bacterial virus is an assembly of
protein components (a). The "head" is a protein membrane, shaped like a
kind of prolate icosahedron with thirty facets and filled with DNA. It is
attached by a neck to a tail consisting of a hollow core surrounded by a con-
tractile sheath and based on a spiked end plate to which six fibers are
attached. The spikes and fibers affix the virus to a bacterial cell wall (b).
The sheath contracts, driving the core through the wall, and viral DNA
enters the cell. (From W. B. Wood and R. S. Edgar, "Building a Bacterial
Virus," Scientific American, July 1967, reproduced by permission.)

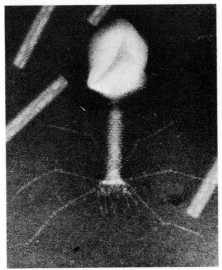

Figure 7. Electron micrograph of a
T4 virus (Courtesy of Robley C.
Williams)

one of the little dots on the fine half-tone reproductions in the *Encyclopaedia*. This, when you demagnify it by 25,000 times, is still 80 angstroms in diameter—32 atoms across, in an ordinary metal. In other words, one of those dots still would contain in its area 1000 atoms. So, each dot can easily be adjusted in size as required by the photoengraving, and there is no question that there is enough room on the head of a pin to put all of the *Encyclopaedia Britannica*.

Engineering at the Top and Bottom

Nature builds marvelous objects on all scales—from the universal to the subatomic. Humans also build marvelous objects, but within a much smaller range of scales. At the top of the range are such objects as skyscrapers, oil tankers, rocket ships, and hydroelectric dams. At the other end, our finest engineering achievements have been in microelectronics, where thousands of individual miniature circuits are integrated on a small, thin slice of silicon (hence the term **integrated circuit**; see Figure 2). But these devices, although marvelous, are primitive in comparison to nature's achievements on the same scale. Ants and mosquitos, for example, are elaborate in structure as well as behavior (Figures 3 and 4). Moreover, there's still room enough for nature to make even ants and mosquitos look gigantic.

A case in point is the T4 virus (Figures 6 and 7). Like other viruses, the T4 needs to grow and replicate within a living cell, in this case bacterial cells. The T4 fastens itself onto the wall of a bacterium, contracts its sheath, and injects a long strand of deoxyribonucleic acid (DNA)—the famous double-helix chain of molecules that encodes genetic information. The T4's DNA resembles the bacterium's own DNA, enough so to trick the bacterium into a fatal mistake: it replicates the T4's DNA along with its own. Soon thereafter the replicated DNA repays its welcome ungraciously by forming into numerous new T4s and rupturing the bacterium's cell wall. The bacterium is destroyed and the new T4s await new victims. The American computer scientist Douglas R. Hofstadter put it well: viral DNA is a molecular Trojan horse.

If the T4 were the same size as the mosquito, I wouldn't be too impressed, but the mosquito is 50,000 times bigger than a T4 virus, and the T4 is one of the larger viruses. The difference is so vast that it's hard to appreciate just how small the T4 is. But consider its size in terms of Feynman's *Encyclopaedia Britannica*, compressed by a factor of 1/25,000 and laid out, page by page, on the head of a single pin. If on top we were to put a single T4 virus, the amount of obscured text would be slightly more than a single letter.

The microscopic world is unfamiliar to most of us. Our inability to see it, touch it, and manipulate it makes it hard for us to accept its reality and to appreciate its potential. But nature has led the way, and we're getting there ourselves. The smallest devices on integrated circuits today are about ten times larger than the much more complicated T4 virus, but they're shrinking. The prospects are exciting.

Computer Printouts and Authority

I'm a skeptical reader; when I read newspapers and books, I'm quick to question their accuracy. But it's a conscious effort. In fact, I approach the printed word with a predisposition to believe. If I pick a book at random from a library's nonfiction section and read that the African spider *Arachnida fallere* has a poisonous bite, my reaction is more to hope that I never run into one than it is to say, "Oh yeah?" and look for a footnote. When I'm personally familiar with something that I read about in *The Washington Post*, I usually notice inaccuracies. But that doesn't stop me from opening the *Post* every morning, fully prepared to believe what I read. Power to the printed.

This power extends to the computer printout. When my checkbook disagrees with my bank statement, I assume that I goofed and I start looking for my error. Some people don't even bother with a monthly reconciliation; they just rely on their bank statement. Of course, most bank's computers have a good track record, and this encourages our reliance. But I think there's more to it than that.

To an extent, I think the urge to believe the printed word comes from the role of the printed word in education. Indeed, since much of what we've all learned came from books, we tend to acquire the habit of belief early in our lives. I also think that the act of publication is seen as mute testimony of accuracy. Considerable effort and expense is often involved, suggesting that someone other than the author has judged the information to be worthy of printing. In a free society accuracy is also encouraged by legal penalties for fraud and libel. Whatever the reasons, the urge to believe the printed word is strong and deep-seated. I had it long before I ever heard of computers or saw a computer printout.

In Machines We Trust

Our industrial tradition is one that glorifies the mechanical. And today, despite the modern vogue for the handmade, our society remains infatuated with machines. Machines transport people, goods, and information. Clocks, traffic lights, and telephones shape our daily patterns. In our factories, in our offices, in our wheat fields, and in our kitchens, we rely gladly on every kind of labor-saving device. We may complain about quality control and service, but we routinely drive cars, board airplanes, turn on microwave ovens, and submit to machines at the dentist. How often have you been on a malfunctioning elevator? How often do you hear about one? Not only do we rely on machines, we have a firmly ingrained habit of trusting them.

This habit, which predisposes us to trust computers, is reinforced by several factors: The computer isn't just a machine, it's a machine that communicates with us. Moreover, it often communicates in terms of intimidating technical jargon, and it can communicate with incredible speed.

Nothing symbolizes our trust in computers better than the printout. I remember how often we stood in front of the glass wall at Yale, staring at the printer. It was exciting to watch it come suddenly to life whenever the computer was ready with results. I think it was the lack of human intervention that made the scene so captivating. And the speed—even in those days the printers spewed results at remarkable rates. Today I'm still drawn to the printer, and I'm not alone. People like to watch computers print answers. And people are prepared to believe those answers.

The computer is a printing automaton. In the computer printout the innate power of the printed word is magnified by our glorification of machines and our trust in them. Machines have always been involved in printing. But before the computer, people didn't perceive machines as participating in the decision of what to print.

In fact, computers don't decide what to print, although we often speak as if they do. Computer printouts are determined by input data and by computer programs. If the input data is wrong, so will be the output—"garbage in, garbage out." Furthermore, correct outputs also depend on correct computer programs, and—as I'll discuss more in Part III—the typical large computer program is considerably more likely to have a major, crash-resulting flaw than is the typical car, airplane, or elevator. The computer may be the ultimate machine, but today it's less trustworthy than many of its predecessors.

Seeing Through the Printout

Feelings of intimidation can be an exaggerated form of respect. In this sense, the typical computer printout is less worthy of your respect than traditionally printed material. A principal reason is that computer printouts are easy to produce and easy to revise. Traditional printing is more difficult—it takes longer, its products are harder to revise, and it usually requires the coordinated activities of several individuals or organizations. These difficulties have encouraged the use of editorial review and other institutional controls that are characteristic of formal publications. Computers can make it easier, quicker, and cheaper to print formal publications; individuals can do what was once practical only for organizations. The computer has reduced the economic and practical importance of institutional controls, and so the controls are often relaxed. Earlier I argued that our predisposition to believe the printed word arises in part from the mute testimony of the publishing act. In the computer age the value of that testimony is disappearing.

Computer printouts are not only getting easier to produce, they're getting harder to recognize. The result isn't always beneficial. One example, which affects me personally, concerns the media that are used to report results of scientific work. Many such results are published in so-called "refereed" journals. These journals are effective as archives, but ineffective as a means of timely communication among scientists—it can easily take years from the submission of a manuscipt to its publication, and by that time the field has moved on.

To stay abreast of current scientific work we depend instead on "preprints"—less formal publications that circulate as carefully typed manuscripts or as technical reports published by a scientist's home institution. Preprints used to be expensive and time-consuming to produce, especially if they contained complicated tables, graphs, and mathematical equations. Consequently, institutional and self-imposed constraints discouraged the indiscriminate publication of preprints. If I received a preprint and knew nothing about its author, at least I knew that it had quite likely been prepared with care; this didn't mean that the preprint was correct, but I would take it more seriously than I would have had it been just a copy of handwritten notes from a laboratory notebook.

Today's computers are transforming preprints. Not only do computers enable a sharp reduction in the "turnaround time" required to produce preprints, they make it easy to produce the preprints. And it's not just easy in terms of the operations that you have to carry out personally, it's easy organizationally—you don't need the support of a publications department, you just need a computer with good word-processing capabilities. Moreover, it's easier for people to comment on your nice-looking drafts and it's easier for you to change them in response to the comments. These are among the advantages of word-processing technology.

But there are disadvantages as well. It has become so easy to prepare professional-looking papers that many people don't bother with the formalities involved in technical reports. Increasingly, I'm annoyed by the slick-looking preprints that I receive. Their contents are worthy only of first drafts, and sloppy ones at that, but they're presented as finished products and they look the part. This divergence of form and content is by no means restricted to scientific papers. You may already have noticed the same phenomenon in your own field as the word processor takes its place besides the Xerox machine, increasing the quantity of papers while decreasing the quality of their contents.

The opposite phenomenon can also occur. There will always be organizations and governments that thrive on the production of false or misleading documents, in which case the economic and institutional controls I mentioned before can operate against accurate publication rather than for it. Here, the computer can cut through the controls, with the positive result of freer and more accurate publications. But it is a double-edged sword that likewise makes it easier to publish misleadingly. As the computer increases the freedom of writers, so does it increase the responsibility of readers.

Computer Anxiety

Our dependence on computers is broad, deep, and intensifying. This fact causes anxiety—anxiety about using computers ourselves, anxiety about dealing with the computers around us, and anxiety about the effects of computers on our society.

Many people want to use computers to improve their personal and professional lives. Some of these people dive easily into the computer mainstream. Others want to get into it, but they hesitate to take the plunge; they're afraid that computers are too difficult, too technical, and too unfamiliar. Still others have these same fears, and they're not at all attracted to computers; unfortunately, they're being pushed.

Many people notice that they get more mail from computers than from people, and they don't like it; they feel victimized by junk mail, credit cards, utility bills, and other examples of computer-controlled commerce. If they disagree with a bill or balance statement, they feel overwhelmed by the effort required to object, and they feel unprepared to do so effectively. They worry about the effects of computers on society. The computer is seen as an adversary.

Computer anxieties are real. They add psychological and socio-logical dimensions to the spread of computer technology.

Keyboard Paralysis

To those who've never experienced it, "keyboard paralysis" sounds like a joke. But it's a real problem for many people, both those who want to use computers and those who'd rather not. The symptoms are obvious: a vague discomfort when asked to sit in front of a computer terminal, a conscious fear when asked to press some keys. The victims ask themselves:

"Will I break it?"

"Will I destroy some information?"

"Will I feel foolish?"

"Will I look stupid?"

"What if it beeps?"

Their hands hesitate; depressing the first key is an act of will. The main reason for this hesitation is the uncertainty of the result. And this uncertainty, this sense of unpredictable consequences, causes anxiety. I've watched people suffer through keyboard paralysis, and I've talked people through it, but only recently did I taste it myself.

It happened when I was shopping for a word processor on which to write this book. For years I've used computers in my work at the Naval Research Laboratory, not just as research tools, but as tools to help me write everything from scientific papers to bureau-cratic memoranda. Indeed, I depend on computers to help me write faster and better. When I decided to write this book, I wanted a personal computer at home to support me in the style to which I had become accustomed. About this time there became available a number of personal computers that essentially were scaled-down versions of the relatively large systems I knew well. This was an exciting development, and, as I began my search, I assumed that I would buy one of them. In the end I didn't. Instead, I bought something different—an Apple Lisa, which was then a brand-new product.

I first saw the Lisa at an hour-long demonstration in a local computer store. I went expecting to leave with a long list of things the Lisa did poorly or not at all, but I left with a short list. I was impressed. About a week later I went to see the Lisa again and to talk more with the people selling it, this time at a "computer show" at the Washington Coliseum. A few days after that I returned to the store and read the Lisa's manuals. The following week I went back again to see a final demonstration and to sign a purchase order. At this point I was well informed about the Lisa, but I hadn't actually used one myself. I had definitely wanted some "hands-on" experience—I couldn't imagine buying a new computer without it— but somehow it hadn't worked out. There was only one Lisa in Washington, and the salespeople didn't exactly encourage every prospective customer to hang around and play. At the final demonstration, however, I had my chance, and of course I took it. The salesperson left the room for something, and without even a furtive glance around, I reached out. But then the unexpected—for a small but acutely self-conscious moment, my hand hesitated. Keyboard paralysis.

It was the mouse's fault, sitting there quietly next to the Lisa. Mouse? I refer not to some unwelcome desktop rodent, but to a palm-sized, box-like object with a button on top, a rolling ball underneath, and a wire connecting it to the Lisa. A wonderful invention, the mouse is used to interact with the Lisa in a relatively natural and efficient manner that supplements use of the keyboard. When you move the mouse around on the desk, a corresponding mark on the display screen moves in concert. As a result, you can use the mouse to point to something and you can then press the button on the mouse as a means of issuing commands.

What bothered me about the mouse? Was it the confrontation with a radical, unfamiliar innovation? Not exactly. Although the mouse wasn't common in personal computers when the Lisa was introduced in 1983 (for example, Apple's Macintosh had not yet been bred), the Lisa is hardly the first computer with a mouse. Indeed, mice—the legitimate plural of a perhaps illegitimate noun— are old by computer standards. Mice were invented about twenty years ago by Douglas Englebart and William English at the Stanford Research Institute (now SRI International) and have been used enthusiastically ever since by many in the R&D community. Nor was the Lisa the first computer to use a mouse in a commercial product that achieved widespread use; that distinction belongs to the Xerox Star. So mice had been around for quite a while. And I had known about them for a while—I knew their history, how they work, what they're used for, and their advantages.

Despite this knowledge, I still hesitated. Why? Simple: I knew all about mice, but I had never used one! Intellectually I knew what to expect, but I didn't have the assurance that comes only with direct experience. In this respect, using a computer is like cooking a soufflé, driving a car, hang gliding, and having sex.

Keyboard paralysis is widespread. The phenomenon is recognized as a significant problem by people who sell computers, train new users, or design training programs. Often it comes with a first— the first time you use a computer, the first time you use a new computer, the first time you use a new computer program. Keyboard paralysis decreases with experience, because experience improves the ability to predict.

Anxiety at Work

The severest cases of computer anxiety occur on the job, where computers are wrenching the status quo. Whether workers face office automation or factory automation, they see their own obsolescence and they're uncertain about their ability to adapt. Also, as more jobs require the use of a computer, fewer of those people who are susceptible to computer anxiety can avoid being exposed. For years, computers were used only by those who were drawn to them and took to them easily. Not anymore.

People are confronting computer technology throughout the work force—shopkeepers, lawyers, secretaries, clerks, accountants, doctors, restaurateurs, assembly-line workers, and managers of all types. Many of them have to deal with computer professionals, but they have no computer training themselves—they feel ill-equipped, intimidated, and insecure. Others are choosing to use computers themselves, or are being pressured into doing so, but they have trouble getting started and they don't get far. Whether or not they suffer initially from keyboard paralysis, once they start typing they become frustrated and annoyed. They keep making mistakes, and they don't understand what they're doing wrong. They have trouble controlling the computer.

A common example of computer anxiety occurs when word processors are first introduced into a traditional office. For secretaries in such a situation, there's considerable pressure to master the new machine. Whether the word processor is seen as an obligation or an opportunity, many are unsure of their ability to master it. What if they fail? Will they lose their jobs? And even if they succeed, will they lose their jobs anyway—not to people who are better-trained, but to a next-generation word processor? These feelings can be intense, with dramatic results.

In 1975 I was responsible for a small group of people engaged in computer science research and development. Around this time word-processing technology had reached the point where it made sense to introduce it into an office such as ours. I proceeded to do so, despite my secretary's implacable skepticism. When the machines arrived, I gave my secretary the responsibility of redesigning the office layout—I asked that the word-processing terminals, the printer, and the other equipment be put in appropriate places— and I left. When I returned, a terminal was near my secretary's desk; but it hadn't displaced the typewriter, which remained, poised for use, on top of the desk. Being a clever technologist but a not-so-clever manager, I switched the positions of the typewriter and the terminal. My secretary retreated, in tears, to the head administrative officer of our division.

I was surprised by this reaction, but I've since found out that it could have been worse. Other reactions to computer anxiety have included hyperventilation, vomiting, and attempts to destroy the offending machine. I was even more surprised by the longevity of the reaction. I knew that there was anxiety and skepticism, but I was confident that the word processor would enthrall my secretary quickly. Wrong again; it took almost a year. To me, the word processor was a predictable, efficient tool that provided elegant solutions to numerous office problems. There were a few "glitches," but these could be explained easily, and they were no more inconvenient than various typewriter-associated inconveniences we lived with comfortably. It was obvious that the word processor would not only make my life easier, but my secretary's as well. To my secretary, however, the word processor wasn't predictable, efficient, and elegant; it was threatening, capricious, opaque, and clumsy.

Anxiety at School

At school, there's pressure for educational computing—teaching about computers and with computers. The pressure comes from parents, students, school administrators, public officials, and some teachers; much of it arises from the assumption that "computing" will soon be a necessary basic skill, just like reading, writing, and arithmetic. Although the importance of "computer literacy" is sometimes overdramatized, it's a valid educational goal. As Robert S. McNamara wrote in 1968:

> A computer does not substitute for judgment any more than a pencil substitutes for literacy. But writing without a pencil is no particular advantage.

Additional pressure arises from the computer's tremendous potential to serve as an effective tool for teaching other subjects. The overall pressure is intense. One indicator is money: educational computing is the one area of the school budget where additional spending is favored by practically everyone concerned.

Spending money for educational computing is one thing; spending it wisely is another. Given the money, most schools proceed happily down the path of least resistance: they buy some computers. They do so without adequate planning for how the computers will be used and without adequate planning for teacher training. Such planning raises tough problems, and there are few readily available, tested solutions. But there's relentless pressure to do *something*, and everyone knows how to go shopping. The results are predictable: intense anxiety for teachers and ineffective educational computing for students. Successful educational computing cannot be achieved merely by placing computers in every classroom, like Gideon Bibles.

During the summer of 1982 I began participating in the planning for an educational computing program at my daughter's elementary school. When I first became involved with the planning group, they were doing the usual thing: surveying the market and deciding what to buy. Thanks to a small group of recalcitrants, we stopped the market survey and began formulating a more comprehensive plan. We did not finish quickly. We wrote position papers and proposals for external funding. We debated endlessly—vocal evidence of the power of anxiety-fueled opinions over facts. Moreover, the debates were among parents, teachers, and administrators—three forceful groups with different concerns and different anxieties. These groups rarely agreed, even among themselves, and progress in the debates, when it occurred, seemed to be random.

A year and a half later we were done. We had an approved plan, not only for buying equipment, but also for teacher training and curriculum modifications. It was a reasonable plan, but it was shamefully modest compared to the time and effort involved in preparing it; if you read it, you might guess it had been produced in a week or so. After another six months, even this plan had been abandoned. Owing to committee fatigue, generous gifts, and a resurgence of the shopping instinct, computers were put into most of the classrooms, while teacher training and curriculum modification were left to evolution.

Our failure attests in part to the difficulty of the problems, to the lack of well-tested solutions, and to the inefficiencies of volunteer committees. But mostly our failure attests to the power of the anxieties involved. I came to realize just how intense and reasonable were the teachers' anxieties. They were being asked not just to accept computers for their own use, and not just to allow their students to use computers—they were being asked to integrate computers thoroughly into their classroom activities. Like professionals in other fields, teachers feel threatened by the technology, and they are uncertain about their ability to master it. But in other fields people need only learn enough to use the technology—teaching requires a deeper level of understanding. Teachers know this, and it multiplies their anxiety. To make matters worse, children are quite without computer anxiety and in many cases already know more about computers than do their teachers.

Anxiety for Society

Computers affect us whether or not we use them ourselves. On the sinister side, people worry not about their ability to master computer technology, but about the ability of computer technology to be used in mastering them. They also worry that the public safety may depend too much on computers. They ask questions about computer crimes, electronic vandalism, invasions of privacy, political power, military power, and the possibility of computer-instigated disasters. Will the biggest robberies of all time be accomplished by breaking into vaults, or by breaking into computers? Will the worst military mistakes of all time be made by generals and admirals, or by computer programmers? Will the worst disruptions of everyday life be caused by natural disasters or by computer failures? Will history record that the most effective manipulations of populations used mass media to broadcast to everyone, or mass memories to record data about everyone?

On the social side, people ask about employment, education, and economics. They worry that computers will take jobs away from people. They wonder if children must learn about computers in order to make it in tomorrow's society. Will computers enlarge the gap between rich and poor? Will the advantages of computers bypass the disadvantaged?

And on the personal side, people ask about their intellects and their relationships. As we depend more and more on computers, will our minds grow dull? As we turn more to electronics, will we turn away from each other?

As consumers, we can choose to participate or not in certain aspects of the computer revolution. But as citizens, our choices are fewer; we feel less in control, a feeling that itself heightens our anxiety about the effects of computers on our society.

Coping with Computer Anxiety

Ralph Waldo Emerson said: "Fear always springs from ignorance." While this is hardly true in general—it would not, for example, be convincing if affixed to my dentist's spotlight—it does apply to keyboard paralysis and related anxieties. Information about how computers work can ease the anxieties associated with using them.

Unfortunately, new users often don't get enough basic information because veterans underestimate how important that information can be. Experience insulates veterans. We take for granted all sorts of skills, intuition, and information; we overlook how often we use them, how hard it can be to acquire them, and how hard it can be without them. Such oversights on my part contributed to my secretary's problems with the word processor. I relied on inadequate manuals, sporadic and sparse explanations, occasional visits by "customer service representatives," and psychological pressure. I should have insisted on a systematic training program.

Information can calm fears about using computers. But fears about the effects of computers on society are a different matter. Some of these fears also spring from ignorance, but many do not. On the contrary, some of them are apparent only to the well-informed. This just means that information is important for a different reason—not to eliminate the fear, although that may happen, but to illuminate the cause.

In both cases, however, information isn't enough. *You* have to start using computers, and *you* have to form your own opinions about the effects of computers. But information can make your actions easier and your opinions educated. Whether you welcome the computer revolution or fear being one of its victims, information can help.

Computers and Sex

Many people recognize that information can help, but they wonder about who can learn. For example, the subject of computer anxiety is tinged with traditional prejudices about the role of women in science and engineering. Arising as a modern blend of mathematics, machinery, and microelectronics, computers arrived with a masculine aura. Traditional stereotypes suggest to some people that computers are harder for women (high tech, don't touch).

Given the legacy of traditional female roles, it could be true that women are more susceptible to computer anxiety than are men. But whatever their initial feelings about computer technology, there's plenty of evidence that women can learn about it, use it, control it, and contribute to it. Large numbers of women operate comfortably at every level of technical expertise in computer technology.

I don't mean only that women use computers. Many women also design, build, program, and study computers. In fact, women have been involved in computer science and engineering from the beginning. An early example was Lady Ada Lovelace. The daughter of Lord Byron, Lady Lovelace was a gifted musician and mathematician. She became fascinated with the "analytical engine," an elaborate mechanical calculator invented by the nineteenth-century British mathematician Charles Babbage. The analytical engine is one of the clearest predecessors of the modern general-purpose computer. Lovelace translated into English an Italian's description of the analytical engine. At Babbage's suggestion she added her own notes, and in doing so tripled the paper's length. She and Babbage worked out a set of instructions for the analytical engine that was similar in many respects to a modern computer program. For this reason some historians have called her the first computer programmer and in 1979 the U.S. Department of Defense's new programming language, ADA, was named in her honor.

Among the first programmers of modern computers was another woman—U.S. Navy Commodore Grace Murray Hopper. (A Navy commodore is approximately equivalent in rank to an army brigadier general.) Commodore Hopper first worked with computers at

Harvard in the early 1940s—in her words, she was "the third programmer on the world's first large-scale digital computer, Mark I." Hopper went on to a distinguished career with the Univac Corporation and the U.S. Navy. She was at the center of many achievements in computer software, including the discovery of the first "bug" (a story that I'll relate in Chapter 2), and including the development of the business-oriented programming language COBOL. Commodore Hopper has received many awards and has had one named after her. Today, seventy-eight years old, she is the oldest U.S. Navy officer, male or female, still serving on active duty. This is not just an example of female longevity—if you're over retirement age, you can't continue to serve unless the Navy still wants you.

In 1980 there were more than 3,000,000 scientists and engineers in the United States, including computer specialists. Only 13 percent of the scientists and engineers were women, but 27 percent of the computer specialists were women. Moreover, the proportion and number of women in computer science and engineering are increasing rapidly: In 1972–73, only 14.9 percent of the bachelor's degrees in computer science were earned by women; in 1980–81, the figure was 32.5 percent.

Participation is one thing, remuneration is another. A 1978 salary survey of "experienced scientists and engineers," defined as those scientists and engineers who were already in the labor force at the time of the 1970 Census, reported that women in the computer specialties had an average salary that was 90 percent of the average male salary. This is *not* equal pay for equal work, but the 90 percent figure is larger than that for any other science and engineering category covered by the survey—including engineering, psychology, life sciences, mathematical sciences, physical sciences, and social sciences.

Why are women more successful in computer science and engineering? The most likely reasons are that computer technology is relatively new, arriving in the 1940s less tinged with prejudice than its predecessors; that women have been involved with computer technology from the beginning; and that the technology exploded from the 1960s to the 1980s, during a time of increasing female participation in the work force and increasing attention to feminist issues. I asked Commodore Hopper for her opinion; she just pointed out that, unlike the jobs of surgeons and lawyers, the programmer and the systems analyst never had a gender-specific connotation.

Women are fully capable of dealing effectively with computer technology. Lots of women are doing it, and 'they're doing it well.

Computers and Age

Perhaps you worry that you're too old to learn about computers and too old to use them. After all, it's often said that the time to learn foreign languages, mathematics, and practically anything else is when you're young. Don't worry.

Consider one example: the world's leaders in the fields of computer science and engineering. Suppose you made a list of the ten most distinguished, respected people working in these fields today, and suppose you looked into the backgrounds of the people on the list. You would find that most of them are, insofar as their computer knowledge is concerned, largely self-taught. Their formal education was in other fields, and most of it occurred before their involvement with computers.

Fifteen years ago it was unusual to meet an accomplished computer user, unless you happened to be "in the business." Today there are millions of accomplished users and it's no surprise to meet one, no matter what business you're in. For most of these users, their first computer experiences took place well past their formative years. They did it, and so can you.

It's Easier Than It Looks

It really is. Most people can understand the basic principles behind computer technology, and most people can learn how to use computers successfully. Many doubt their ability to succeed because their anxieties interfere with learning and make them overestimate the difficulties involved. This is a common and natural reaction; it can happen when you're confronted with any new tool—a new car, a new camera, a new sewing machine—but it's a debilitating reaction only if you let it be one. The reaction is due more to unfamiliarity than to intrinsic difficulty.

Unfortunately, despite claims that today's computers are "user-friendly," learning to use them effectively often requires some technical knowledge and intuition on the part of the user. Such requirements are due mostly to equipment limitations and poorly designed software. More capable equipment and better-designed software are steadily becoming available, and this trend will continue, but it's still easy for newcomers to feel intimidated and to overestimate the difficulties involved. But a little knowledge of computer jargon, a little knowledge about computer technology, and a little insight into the nature of the design failures and equipment limitations that lead to unpleasant computer systems—all of these are easy to come by and all of them go a long way in helping the newcomer to cope.

Although the technology and its accompanying jargon are new, many of the underlying concepts are neither new nor hard to understand. As Tallulah Bankhead is said to have remarked to a companion while attending a theater performance:

"There's less in this than meets the eye."

Why Bother?

Overcoming computer anxiety and learning more about computer technology has numerous benefits, both for ourselves and for our society.

Personal Benefits

At work, overcoming computer anxiety can help to raise morale, reduce disruptive tensions, improve performance, improve productivity, and encourage both people and organizations to achieve new goals. At home, computers can improve our lives in a variety of ways. At school, computers can help us achieve new insights as we learn faster and more thoroughly. Computers can also be fun, whether or not they are used to play games (a relatively recent emphasis).

There is an undocumented but much discussed advantage that

many believe to result from using computers and writing computer programs. The most popular term for this advantage seems to be "procedural thinking," and it refers loosely to supposed improvements in logical thinking and problem solving. I don't like the term —a better one would be "disciplined thinking." But although this effect is harder to achieve than is commonly supposed, I believe that it's real. The study of computers and computer programming can lead to improvements in your ability to analyze and solve problems.

Political and Social Benefits

The political and social costs of computer anxiety are as subtle as they are important. For example, despite glib references to computer errors, many people are intimidated not just by computers, but by computer outputs. Given the extent to which commercial and governmental communications are dispensed by computers (just look in your mailbox), such intimidation is bad. It discourages people from questioning authority, surely one of the most dangerous effects any technology can have. Learning more about computers can make it easier for people to meet them head on, and easier to go around them.

Learning more about computers can also help people to participate intelligently in the formulation of public policy. This is important because the long-term effects of computer technology will depend less on the technology itself than on society's response to it. Powerful technologies have powerful effects, and the potential effects of computers are both good and bad. We all share the responsibility of defining and achieving an appropriate balance.

PART TWO

The computer is an information-processing machine. To use one effectively once required detailed technical knowledge, but technical improvements have steadily reduced this requirement and made computers much easier to use. This is a welcome trend, but it has not progressed nearly as far as users want and advertisements claim. Today, ease-of-use is rarely a constant experience.

Fortunately, a little technical insight can make it easy to get over the rough spots, and sufficient insight is easy to obtain. The computer's logic is obscured by miniaturization and cloaked in jargon, but the basic principles are simple. Most people can understand how computers work; they can understand what the different components of a computer system do, and they can understand how those components function together to process information and interact with users.

Computer jargon is one of the biggest stumbling blocks, but it's also one of the easiest to remove. Part II therefore begins with two chapters on computer jargon. The first contains a general discussion of the purposes and the effects of computer jargon, both good and bad, while the second contains brief definitions of key computer components and concepts. Next, Chapters 4 to 6 focus on one of the most important and relevant aspects of computer technology: how computers interact with their users. I explain the nature of these interactions, the ways in which they affect us, and the distinguishing features of computer systems that interact effectively. I also explain why computers converse with us in the ways that they do, and why our coversations with computers are often more difficult and frustrating than conversations with people. In Chapter 7, which ends Part II, I go beyond our interactions with computers to the underlying software, and I explain what computers are doing when they process information and converse with us.

The trend toward ease-of-use will continue, and technical knowledge will continue to become less important. Meanwhile, some technical insight can help you to overcome computer anxieties; it can help you to choose a computer system that works for you, and it can help you to use that system effectively. Moreover, it will always be useful to understand the basic principles. Whenever you understand the logic of a machine, it's easier to use that machine effectively.

CHAPTER 2

Coming to Terms
with Computer Jargon

Computer jargon is pervasive and annoying. When people complain about it, they are reacting to the use of technical terms not generally understood, to the impenetrable style in which much documentation about computer systems is written, and to the widespread use of computer slang.

Computer jargon makes computers harder to approach, and it contributes to the frustrations and anxieties of novice users. It is, however, unavoidable. Whether you want to use computers comfortably or just to discuss them intelligently, coming to terms with some jargon is a necessary step.

Jargon and Anxiety

Whether people choose to use a computer or are forced toward one, their goal is usually to use the computer as a tool. They need to analyze a budget, draw a diagram, or write a letter, and they know that the computer might make the job easier. Often their first experience is negative. They conclude quickly that in order to write a letter they will have to learn a lot about how computers work.

One of the reasons for this reaction is jargon. People want to deal in terms of paragraphs, tab stops, and spacing—terminology and concepts relevant to the job they are trying to do. But they find they also have to deal in terms of disks, RAMs, and warm boots (a verb, it turns out). Not only does this often-unnecessary jargon distract people from the job they're trying to do, it reinforces their suspicion that they can't do the job without learning all about computers. The jargon can make them feel stupid, or at least that they should know more.

Users expect to make mistakes, and they do. But instead of being told

YOU CAN'T PUT A PARAGRAPH THERE . . . TYPE ? FOR HELP

they're told

DOS SYNTAX ERROR

(often accompanied by an accusatory beep). The intimidating effect of the technical terms is compounded by the clipped style. Instead of conveying the attitude "People make mistakes, and computer programs have limitations," the style conveys the attitude "You should have known better." This just feeds people's anxieties.

Computer programmers also make mistakes. And electronic components sometimes fail. But when such events cause problems, it rarely helps the new user to be told

FATAL ERROR . . . ILLEGAL OPCODE AT B372

or

MEMORY VIOLATION . . . PARITY ERROR AT A3F2

The jargon in these messages is particularly debilitating, since at the same time users are likely to discover that their previous work has been lost and that the system no longer behaves in a customary manner. Because such messages usually appear immediately after a user has typed in something, users tend to blame themselves, a tendency exacerbated by the unfamiliar jargon and the accusatory tone. In a pedantic sense, the user *is* the cause, but the fault lies elsewhere—either with physical breakage or, most likely, with the person who wrote the computer program.

Users trying to write letters don't care about DOS, illegal opcodes, and memory parity. And they shouldn't have to, because these things are not properly the concern of someone writing a letter. In general, jargon-filled error messages are symptoms of a basic problem, namely that the designers and programmers of many office and personal computer systems have failed to separate their own concerns from those of the user.

The Language of Death and War

Not long ago, my eight-year-old daughter, Hilary, and I were sitting together, using a computer. At one point she asked me whether we could do something in particular. I pointed to the name of a program and said that we could do what she wanted by having the computer execute that program. Hilary turned to me, all eyes. "Execute?"

Of course, we say "execute" in the sense of carrying out, or putting into effect, but that's not always obvious to the uninitiated. I'm not sure why we "execute" programs rather than "process" or "operate" them (sometimes we "run" them). It appears that the terminology arose in the early days of electronic computing, when computer programs were called "order codes" and it was, in a sense, consistent to say that the computer executed orders. But even though we say "execute" in the nonviolent sense, violent language crops up a lot in computer jargon.

On occasion, one of the computers in my branch at the Naval Research Laboratory stops working. If I wander out of my office and say: "Hey Alan, what happened?" he might reply: "Your last hack crashed the system. One of your processes branched to an illegal address, tried to execute Joe's code, and died after committing a fatal protection violation. The broken pipe killed seven more processes, hung two others in a deadly embrace, and eventually caused the system to bomb with a core dump."

Language like this encourages the anxious computer novice about as much as medical humor encourages the preoperative patient. Novice users often worry about doing something wrong that will damage the computer or destroy information. After a hesitant keystroke, a message like

ILLEGAL INPUT . . . FATAL ERROR . . . JOB KILLED

can easily be fatal to their interest as well. No wonder. They can only judge the meaning of this language by previous experience, which hardly conditions them to conclude merely that the computer program stopped after processing something unexpected. In fact, unlike the same language in "real-world" contexts, the offending message is mostly computer slang that reflects accurately neither the user's error nor its consequences. The sooner the novice user understands this the better.

Computer jargon may not be more violent than language in other human endeavors, but there's a major difference: unlike violent language in other contexts, violent computer jargon doesn't reflect actual violence. Violent computer jargon is used to describe an artificial environment, within the computer, that actually is free of violence and physical danger.

Does the violent jargon reflect some basic psychological problem of the glassy-eyed, nocturnal computer freak? (I can already see the thesis title: "Alienation and the Computer Nerd: Metaphors of Death and War in Modern Computer Science.") I doubt it. I suspect that it's just human nature to act out violent impulses in a perfectly safe environment. This may also explain the widespread popularity of arcade and personal computer games, most of which are based on violent themes.

Most computer users seem to like the violent slang; once they understand it, newcomers take to it with relish, especially when they realize that the violent slang is used with a sense of humor rather than malice. For example, I often use a particular program to display electronic mail that arrives from computers all around the country. Each message starts with a few lines that show when and from where it was first sent. Additional lines show how the message was routed between the originating computer and ours. All of this information comprises the message "header," which can be quite long. Long headers are annoying because they clutter up the computer output without conveying much useful information. If you read the messages on a slow computer terminal, long headers also waste time. Accordingly, the mail-processing program has an option to strip off the headers before displaying the messages. In the program documentation this is referred to as "Marie Antoinette mode."

Vocabulary and Communication

Like professionals in other fields, computer professionals need tools for precise communication. A technical vocabulary is one such tool, and it's a necessary one. Since there are occasions that require precise communication not only among professionals, but also between professionals and nonprofessionals, nonprofessionals must expect to confront this vocabulary to some extent.

It is ironic that, although a technical vocabulary's purpose is to facilitate precise communication, it easily can have the opposite effect. The problem is that the professional, who may be thinking clearly about *what* to communicate, isn't thinking clearly about *how* to communicate. Often the jargon is just a habit; it helps the expert, who assumes that it helps everyone. (The road to Gehenna is paved with good intentions.)

In choosing words for a technical vocabulary, there are two approaches: either the words are coined specially for use within the vocabulary, or they are taken over from the general vocabulary and assigned a restricted or special meaning. A newly coined word can be a necessity—after all, technology evolves faster than language—but the general tendency is to take the second approach whenever possible. There aren't too many coined words in computer jargon—"software" is perhaps the best example. In general, coined words don't cause major communication problems, provided that they are used judiciously and defined clearly. If you encounter such a word unexpectedly, at least you recognize that you don't know it, and you can look it up.

Words taken over from the general vocabulary or from other fields are more of a problem. They can impede communication because you may think you know what they mean. Their technical meanings usually share a lot with their common meanings, but the differences are important. Some examples from computer jargon are "compile," "function," "word," "buffer," "file," and "index." Experts are so accustomed to the technical meanings of such words that they can easily forget how the meanings are clouded by conventional usage. Some words from computer jargon have, in addition, a technical meaning in another field. These words can be a particular problem if you happen to be a professional from that other field, used to interpreting words in a restricted, technical sense. Some examples are "procedure," "operation," and "expression," which have medical as well as computer interpretations.

The common word "bit" is in between these extremes—it is both a technical neologism and a word in the noncomputer vocabulary. A **bit** is the smallest unit of information stored in and processed by most computers. The word is a neologism derived from mathematical terminology (**binary digit**—more of that later), but it is also a pun on the common word used to mean "morsel."

Difficult Documentation

Like other complicated machines, computers and computer programs are documented by publications that describe their specifications, design, construction, internal operation, and use. Such publications are referred to collectively as **documentation.**

Most people care only about **user documentation**—documentation intended for those who use a particular computer or particular computer program; the general term "documentation" is often used in this restricted sense. Although they are sometimes combined into a single document, there are two principal kinds of user documentation: **user manuals,** or **tutorials,** which are designed to teach someone how to use a computer or computer program, and **reference manuals,** which are designed for looking up specific facts. When the computer itself is used to present documentation, the documentation is referred to as being **on-line.**

Bad documentation is one of the biggest complaints of all computer users and one of the biggest impediments for novice users. Erroneous documentation is the worst offender, but the most frequent is documentation that's hard to read and hard to use. This occurs when documentation is poorly written, laced with unnecessary jargon, and overloaded with irrelevant technical details. It also occurs when the documentation reflects accurately some poorly designed hardware or software. But good documentation is hard to write, even for a well-designed system; understanding why this is so can help the new user.

Computerese vs. Legalese

For a long time I disliked legal language, which struck me as confusing and impenetrable. I thought its purpose was to clarify relationships, but its nature seemed to interfere with this purpose.

Some years ago my feelings about language in general underwent a sort of stylistic conversion: I became a card-carrying crusader for stylistic simplicity. (Had there been a Strunk and White ashram, I would have been there.) The effect on my feelings about legal language was instantaneous: not only did I dislike it even more, I became self-righteous about it. Fortunately, my "religious studies" soon led me to the writings of Sir Ernest Gowers, whose book *The Complete Plain Words* contains an eloquent chapter on the subject. In "A Digression on Legal English," Gowers set me straight. Here is a summary of his arguments.

The most commonly seen forms of legal writing occur in documents that concern the rights and obligations of individuals, organizations, and governments. But the purpose of these documents is not to describe those rights and obligations; it is to define them. It follows that the paramount goal of such legal writing is not to be readily intelligible; it is to be unambiguous. In writing a legal document, therefore, one must guard against every conceivable interpretation that conflicts with the intended rights and obligations. In part this is achieved by the seemingly endless enumeration and delimitation of possibilities. In part it is achieved by the stilted avoidance of pronouns, so that questions of antecedents never arise. And in part it is achieved by relying on those hackneyed phrases and constructions that make legal language such an easy target for satire; because it is the courts that must resolve disputes about the meaning of the chosen language, it's best to stay as much as possible on the firm ground of established interpretations. As Gowers put it: "No one can expect pretty writing from anyone thus burdened."

Legal writing reminds me of computer documentation. Here's an example, taken more or less at random from a user manual for a popular personal computer:

> The buffer name is derived from the name of the last file from which a spreadsheet has been copied. Specifically, it consists of the first eight characters of the filename, disregarding both disk drive designation, and suffix. . . . If the spreadsheet is new and still bears the default name "DEFAULT.PC" (see page 19), the buffer is simply designated the "default" buffer. Should the spreadsheet be stored using the WRITE FILE command (see Chapter X, page 203), or should another spreadsheet be read into this default buffer using the READ FILE command (see Chapter XI, page 220), the filename will change to reflect the new filename. The buffer will remain the "default" buffer, however.

Computer experts read documentation like this easily, just as lawyers read contracts easily. But to novice computer users, such documentation seems about as helpful in using computers as the tax laws are helpful in filing tax returns. Like trouble with legal documents, trouble with computer documentation arises from the prevalence of technical terminology and slang. But trouble also arises from another similarity between legal writing and computer documentation—the importance of being unambiguous.

Pick a Number, Any Number

If I say to you at the start of a magic trick, "Please pick a number from one to ten, write it on a piece of paper, and show it to me," that will usually be sufficient information for you to carry out the task correctly. I did not, for example, have to define for you the meaning of "number." Moreover, whether you choose to write "seven" or "7" or even "VII," where on the paper you choose to write it, and whether or not you choose to circle the number or perhaps draw a box around it—none of these choices will interfere with my knowing which number you picked. It is unlikely that you would write "7.1," but if you did I would say something like, "I meant for you to pick a whole number; please do it again," and that in most cases would be enough to make the second try a success. Similarly, if you wrote "0," I might say, "Zero isn't allowed; please pick a number between one and ten." All of this is quite standard and easy for both of us.

If I were replaced by a computer, however, this interaction would be harder. Even in the restricted context of this example, most computers—and especially personal computers—are not nearly as flexible, as forgiving, or as helpful as I am. Indeed, one of the first things any new computer user notices is the highly structured and exacting nature of the interaction. It is this inflexibility that makes it so important for computer documentation to be unambiguous.

Consider how a computer might have started the same magic trick: It would begin by displaying the message

PLEASE ENTER A NUMBER BETWEEN 1 AND 10:.

You respond by pressing keys on a typewriter-like keyboard. (You may have noticed that I've been using a sans-serif type face [Helvetica] to indicate messages typed or displayed by a computer. I'll do so throughout the book, and I'll use the same convention to indicate inputs typed or otherwise entered by a computer user. To avoid ambiguities or to emphasize exactly which characters are typed, I often enclose the input text in single quotes when it is embedded in a sentence rather than displayed as a separate line.)

If you type 'SEVEN' or 'VII' in response to the above message— that is, if you type the characters within the single quotes—the response might be a shrill beep together with the displayed message

ILLEGAL NUMBER, TRY AGAIN:

or

FORMAT ERROR

or

SYNTAX ERROR . . . ILLEGAL TERMINAL SYMBOL

or perhaps just

WHAT?

The computer might just beep without giving any message at all. If you type '0' (zero) instead of 'SEVEN' or 'VII', the computer might display the message

THANK YOU. NOW WATCH THIS TRICK . . . ,

appear to sit idle for some time, and then announce

FATAL ERROR . . . REGISTER OVERFLOW AT AF45
712 547 234 232
777 234 342 455
209 487 439 332
≥

This sort of behavior is so common that experienced computer users deal with it easily. But not the novice. Quite apart from the intimidating terminology and tone (or beep!), the novice user simply doesn't have enough information to interpret these messages, which makes it hard to see the magic trick.

The problem, of course, concerns what the computer will accept as a number. If you were to type '7' instead of 'SEVEN' or 'VII', all would be well and you would see the trick, but how are you supposed to know that? Presumably by reading the documentation, which should contain an unambiguous description of how to type in a number. But to write such a description is not as easy as it sounds.

What's a Number?

Figure 8 shows how numbers are defined in a user manual for a popular computer program. Except for the name of the program, which I changed to Divine Calc, the example shows exactly what appears on one page of the manual. Please read it now.

I've shown this description to quite a few people, most of whom at first felt it to be complete and relatively clear. But is it? What is the correct way to type in the number "four tenths"? May you type '.4', or must you type '0.4'? The description above the box in

How does Divine Calc recognize a number?

When you begin typing a number, Divine Calc looks at the first character and attempts to decide whether or not what you are entering is, in fact, a number. To be a number the **first** character must be either:

- a hyphen (i.e., minus sign): " − "
- a period (i.e., decimal point): "."
  or
- Any digit: 0–9.

Once Divine Calc has determined that you are indeed entering a number, it watches to see that the number follows certain rules.

Specifically, a number can have, **in the following order:**

A minus sign (optional)
followed by
Zero or one or more digits
followed by
A decimal point (optional)
followed by
Zero or one or more digits
followed by
An 'e' or 'E' (optional exponent sign)
followed by
A ' + ' or '–' (optional)
followed by
One or two digits of the exponent

Figure 8. One manual's definition of a number.

Figure 8 states that the first character can be a decimal point, which suggests that '.4' is OK, but the description within the box states that the optional minus sign is always followed by "Zero or one or more digits." Does the "Zero" in this phrase refer to the digit "0"? If so, then "Zero or one or more digits" means that you must at least type the digit "0", which in turn implies that you have to type '0.4' instead of '.4'. Or does the "Zero" mean that zero digits (i.e., no digits at all) can appear after the optional minus sign and before the optional decimal point? This interpretation implies that '.4' is OK, which is consistent with the earlier description of what the first character must be. But if this is the correct interpretation, why does the description use the phrase "Zero or one or more digits" instead of "Zero or more digits" or "One or more digits (optional)"?

What about the number '5E–2'. What happens if you leave out the 'E' and type '5–2'? This appears to be a legal number—the description in the box allows an optional '–' after an optional 'E', but it doesn't require that the 'E' be present if the '–' is present. For that matter, which key do you press for the optional '–'? If it is the same key as the one for the optional minus sign that can begin a number, why does the description at the top of the box refer to "A minus sign (optional)" rather than to a '–'?

Enough.

The user's need for complete, clear, and unambiguous documentation is not easily met. This brief example shows how one small aspect of using a computer program can require a lot of discussion, how hard it is to be unambiguous, and how easy it is to overlook special cases.

Descriptions vs. Specifications

There is one kind of computer documentation for which the analogy with legal English is especially apt: specifications. A **specification** is a statement of requirements—a computer program specification, for example, states what a particular program is required to do. The technical meaning of the term is taken from engineering, a field in which it is traditional to specify, in advance of building something, the properties we require of it.

In contrast to a user manual, which describes the behavior that a computer program happens to have, a specification for that program defines the behavior that the program is required to have. If the program's behavior disagrees with the user manual, then the manual is wrong; but if the program's behavior disagrees with the specification, then the program is wrong. Being readily intelligible is less important for specifications than being complete and unambiguous, which further illustrates the analogy with legal English. Indeed, a specification for two computer programs that interact with each other is, among other things, a contract that defines their mutual rights and obligations.

Like legal documents, specifications are hard to read, especially since they may contain complicated mathematical equations in addition to complicated English sentences. But, just as legal documents are intended to be read primarily by lawyers, computer program specifications are intended to be read primarily by programmers. Program specifications tend not to be useful as documentation for the program user, and they are rarely provided as such. Unfortunately, user documentation for a program is often created by attempting to translate its specifications quickly into simple English, which is probably how Figure 8 was created. As Figure 8 illustrates, the result is usually a document that's both inaccurate and hard to read.

There's No Excuse

The importance of being unambiguous doesn't excuse user documentation from being readily intelligible, although it does help to explain why good computer documentation is hard to write. Unlike most legal writing, and unlike specifications, the paramount goal of user documentation is not to be unambiguous; it is to be helpful. Whether or not the documentation is complete and unambiguous, if it's too hard to understand, the user will either ignore it and experiment with the computer, or ignore it and the computer both. This means that the writer cannot ignore readability while seeking to avoid ambiguity. This is an additional burden, one not carried by those who write legal documents.

Fortunately, those who write computer documentation have several advantages. For one thing, the behavior of a computer program can be described in terms of well-defined events and sequences of such events. For another, if the program was designed well, there is a well-understood, logical structure on which to base the description. Of course these things are not true for every computer program, but they are likely to be true for any program worthy of general use.

Another difference is that those who write computer documentation can exploit a wider range of language than can those who write legal documents. For example, computer documentation doesn't have to adhere strictly to precedent-laden terms and phrases. It can also range freely beyond English prose—exploiting diagrams, pictures, and the language of mathematics.

Many novice computer users accept inadequate documentation readily. Perhaps they assume that the inadequacy is theirs. Perhaps they assume that good documentation is an unreasonable expectation. Perhaps their outrage has already been spent on IRS forms. But, while there are many explanations for inadequate documentation, there are no good excuses. Inadequate documentation abounds not because it is necessary, but because it is tolerated.

Anthropomorphic Language

Behind a working computer are people. People are responsible for building the computer and for programming it. But these facts are sometimes obscured by language implying that the computer has a life and a personality of its own. When we say

"It told me to enter a number,"

we mean

"It was programmed to ask for a number."

When we say

"It doesn't know how to do that,"

we mean

"It hasn't been programmed to do that"

or perhaps

"It hasn't sufficient capacity to do that."

When we say

"The damn thing bombed and destroyed my document when I tried to print it out!,"

we mean

"The programmer screwed up!"

or perhaps

"What a time for the electronic equipment to fail!"

We also exploit anthropomorphic analogies to explain how computers (electronic brains) work.

When we talk and write about computers in an anthropomorphic way, perhaps we are just taking linguistic shortcuts. Or perhaps we are just taking advantage of useful analogies. But I think there's more to it than that. We tend naturally to animate objects by the language we choose in talking about them, and we derive emotional satisfaction from doing so. Although the tendency is widespread, it has special force in the case of computers because of their interactive nature. Because they appear to speak for themselves, computers reinforce our anthropomorphic attitude toward them.

Novice users tend to regard computers as having Delphic wisdom and other intimidating, superhuman capabilities. Anthropomorphic attitudes reinforce this tendency, and they encourage people to misunderstand computers and computer programming. For example, anthropomorphic attitudes distract us from the real cause of failure. When we encounter some particularly noxious flaw, we get angry not at the designer or the programmer, but at the computer! This makes as much sense as shooting the messenger who brings bad news. The effect is harmful because it encourages us to tolerate bad design.

The Pronoun Problem

Opportunities to encourage or suppress anthropomorphic attitudes arise constantly in **tutorial programs**—interactive computer programs that are designed primarily to educate the user. Tutorials are programmed to explain new material to the user, to ask the user to perform simple tasks that depend on this material, and then to provide helpful feedback. Tutorial programs are often used as a form of on-line user documentation, and they are also used to teach many subjects, from arithmetic to zoology.

A tutorial program might begin by having you type in your name, after which you see a message like

Hi John, I'm going to help you learn touch typing,

or

Hi John, we're going to learn touch typing together,

or perhaps

Hi John, we're going to teach you touch typing.

To whom do the pronouns refer? Another possibility is

Hi John, my name is Adele Goldberg. I wrote this computer program to help you and others learn touch typing.

The differences among these messages are subtle, but significant in terms of anthropomorphic implications. I prefer messages like the last one, which depersonalizes the computer and emphasizes that programming is a human activity.

The pronoun problem is most significant for tutorial programs, but it arises for any interactive program. One example concerns the computer that I'm using to write this book: the Lisa. Notice that I said "the Lisa" and not "Lisa." The first is the name of an object; the second is the name of a person. All the documentation for the Lisa sticks carefully to the inanimate form of reference, as do all the messages that are displayed to users. A typical message starts with

The Lisa is having technical difficulties . . . ,

and not with

Lisa is having technical difficulties. . . .

How should messages issued from an interactive program refer to the program, to the computer, and to the programmer? Stated differently, whom should the user perceive as the narrator—the computer, the computer program, or the author of the program? For books, an analogous question would be whether or not the reader should perceive the book as the narrator, but it never arises because books aren't interactive. It is a new question in literature.

It's Hard to Resist

Like many other people, after learning to write computer programs I immediately went into "Frankenstein mode." This entails arranging for sardonic output messages that make the computer seem like a person, preferably intelligent. One can explore this genre artfully, and it's great fun. The tendency to anthropomorphize computers is clearly a strong one, and it's hard to resist.

Should we resist? Yes and no. Language that encourages anthropomorphic attitudes about computers can be harmful, and we should take pains to avoid its harmful effects. But to eschew it completely isn't realistic, and even if it were, the results would be dull and cumbersome. For every phrase like "the computer responded by . . . ,"

we would have to substitute

"the computer was programmed to respond by. . . . "

To be thus condemned to the purgatory of the passive voice would be unbearable.

Moreover, provided we don't take them too literally, anthropomorphic analogies can be helpful. People often learn easily about new things by fitting them into models they already have well in hand. Gentle introductions that proceed by analogy work this way, and they can be as helpful with computers as they are with other subjects.

Social Purposes of Computer Jargon

Technical jargon is a tool for precise communication, but it serves some social purposes as well. As a private language for those within a technical field, jargon provides both security and enjoyment. Indeed, these social needs are strong enough to result in computer jargon that serves no real technical purpose.

Protecting the Turf

Technical jargon in any field is a barrier to outsiders. This barrier may reflect legitimate requirements for education or apprenticeship, but it also provides a sense of belonging to the cognoscenti inside. Every profession has its private-club aspects, and jargon is one of them. It feels good to be in the know.

The barrier of technical jargon can also serve an economic purpose. Every professional field is a knowledge monopoly, and in many cases a monopoly has an economic advantage. Jargon helps to protect that advantage. It makes it harder for outsiders to practice the profession, it encourages those who need the knowledge to get it from professionals, and it encourages the belief that the knowledge is worth paying for.

Hangar Flying and Social Hacking

Flying airplanes was once my favorite hobby. Flying is something you do more by yourself than with other people; even when others are involved, it's not a social activity. For one thing, you're supposed to concentrate on flying. For another, the right stuff requires a taciturn demeanor. Besides, the planes I flew were small and noisy. But all this changes when pilots are on the ground and planes are in the hangar. Once out of their planes, pilots congregate and talk endlessly about flying. This is hangar flying.

Flying is expensive. But hangar flying is free, and it's almost as much fun as flying. Indeed, like other pleasurable activities, flying itself wouldn't be nearly as much fun if you couldn't talk about it afterward; hangar flying fills social and emotional needs not met directly by flying itself. Moreover, flying jargon is an essential ingredient of hangar flying. The pleasure and sense of belonging simply wouldn't be there without flaps, trim, instrument approaches, ceilings, DFs, VORs, and radials (not tires).

Computing is not so different. Just as there are flying freaks, there are computer freaks. The slang term is hacker. The **hacker** is a dedicated and enthusiastic programmer, one who revels in marathon programming sessions that exploit every feature of a computer system, including features not intended or even known by the designers. Lately the meaning of the term "hacker" has become distorted as the media have lionized young hackers who have cracked the security of computer systems around the country. Hacking is becoming identified with such electronic vandalism. True hackers—i.e., hackers in the original sense of the term—are outraged by this change in meaning. Pedantry hasn't exactly been part of their image, so it's been amusing to observe their outrage flashing electronically around the country's computer networks. But I side with the pedants; the original meaning of "hacker" is useful and worth preserving. As for the youthful vandals who cruise our electronic highways and byways looking for systems to crack, I call them **punk hackers.**

Like flying, hacking isn't a social activity. Indeed, hackers spend far more time alone at the computer terminal than most pilots spend in the cockpit. And, like pilots, hackers love to talk about their exploits; they thrive on the computer equivalent of hangar flying—I call it **social hacking.** There's lots of slang used in social hacking, even more so than in hangar flying, perhaps because the computer field is so new and different that it invites the invention of terminology and slang. Hackers do this with abandon.

You don't have to be a hacker to enjoy social hacking. Indeed, from novices to experts, most computer users enjoy talking about their experiences. And for all these people the jargon is part of the fun. Consider just one example: the word "bug." In computer jargon, a bug is neither an insect of the order *Hemiptera* nor a modern instrument of covert surveillance—a **bug** is an error. It is a mistake made by a computer programmer, a computer designer, or a computer builder.

We could say "error" or "mistake," but it's much more fun to say "bug." Moreover, "error" and "mistake" capture nothing of the sense of chase. In computers as in nature, bugs are elusive and illusory. Many are small and hard to find. Just when you think you've got one, you find it's not where you thought it was. And just when you think you've got the last one, another shows up. Exterminating bugs—**debugging** in computer jargon—is a lengthy, difficult, but extremely satisfying process.

Besides, behind our use of the term is a good story. It comes from Navy Commodore Grace Murray Hopper, whom I mentioned in Chapter 1. In 1945 she was a programmer on the staff of the Mark II, a pre-electronic computer whose components were electromechanical relays—metal switches that can be forced back and forth between two positions by means of electrically controlled magnets. On one occasion the Mark II stopped, and the staff identified the cause as a failed relay. They examined the relay. As Commodore Hopper put it:

Inside the relay—and these were large relays—was a moth that had been beaten to death by the relay. . . . Now, Commander Howard Aiken had a habit of coming into the room and saying, "Are you making any numbers?" We had to have an excuse when we weren't making any numbers. From then on if we weren't making any numbers, we told him that we were debugging the computer. To the best of my knowledge that's where it started.

And so, even though we use the term today in a different sense, the first bug was one of nature's own. You can see its picture in Figure 9.

Some Advice for the Novice User

Computer jargon can lead to anxiety, confusion, information, and fun. To an extent, the balance among these possibilities is up to you. If you are a novice user, one of the easiest and most helpful things you can do is to learn a little vocabulary. Get a good computer glossary and look up words you don't know. Watch out for words you think you know, but are used so frequently that you suspect them to have a special meaning; look them up, too.

In selecting computers and computer programs, one of the most helpful things you can do is read the documentation. No matter how many things you are told that the computer can do, no matter how many things you see it do, the quality of the documentation is a reliable indicator of how useful the system will be. When you listen to the salesperson and when you read the documentation, remember which of the two you are more likely to take home.

To eliminate a source of anxiety and to facilitate communication, learn some computer jargon. If only this were enough! But communication depends also on the transmitter, and a carefully tuned receiver is helpless in front of a sloppily tuned transmitter. Using technical vocabulary and communicating with it are not the same thing, and it is a sad fact that technical vocabulary is used carelessly in much of the computer documentation that you're likely to read. Moreover, verbose, bureaucratic-sounding prose is as common in the computer field as in any other. Recently I looked at a manual

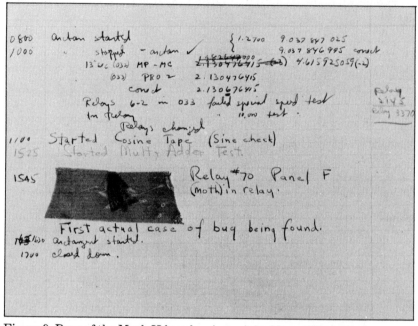

0800   antan started
1000   "   stopped - antan ✓   { 1.2700   9.037 847 025
        13° uc (032) MP - MC   1.9825409000   9.037 846 995 conset
                                2.130476415 (-3)   4.615925059(-2)
        (033)   PRO 2   2.130476415
                conset   2.130676415
        Relays 6-2 in 033 failed special speed test
        in relay   "   10,000 test .
                Relays changed
1100   Started Cosine Tape (Sine check)
1525   Started Multy Adder Test.

1545   Relay #70 Panel F
                                (moth) in relay.

        First actual case of bug being found.
1630 antangent started.
1700 closed down .

Relay
2145
Relay 3370

for CP/M—a popular control program for personal computers. In CP/M, it seems, you don't find programs that can be run or even executed; instead you find "primitive operations which can be programatically accessed." Unfortunately, you cannot relax as soon as you've learned some computer jargon. You will have to be not only an attentive reader, but a discriminating one.

CHAPTER 3

A Small Dose
of Terminology

Learning some vocabulary is one antidote to computer anxiety. Here's a small dose. I'll be using most of it in later chapters.

Hardware, Software, and Systems

For our purposes, most modern computers can be described in terms of three physical units: central processing unit (CPU), memory, and input/output (I/O). These components are referred to collectively as **hardware,** a reference both to their physical reality and to the difficulties inherent in changing them. In contrast, and for opposite reasons, computer programs and their data are referred to as **software.** Usually there is a single CPU, which may have some memory built into it, and one or more additional memory and I/O units. The hardware for memory and I/O components is sometimes contained in physically separate units that plug into the CPU; such external components are referred to collectively as **peripherals.** The term **computer system** refers to an operating combination of CPU, memory, peripherals, and software, although it is common just to say "computer" or "system."

A **computer program** is a sequence of **instructions,** such as "add two numbers," "subtract two numbers," "skip the next instruction if two symbols are identical," etc. Such instructions are stored in the **memory** and acted upon by the **central processing unit (CPU).** The memory is a device that records and retrieves symbols (often numbers), and the CPU is a device that transforms symbols in ways specified by the instructions. The **data** comprise the information on which the instructions operate (for example, the "two numbers" mentioned above), and also are stored in the memory. Locations in memory are identified by numerical **addresses.** When the CPU retrieves information from a specific address, we say that it **reads** that location. Similarly, when it stores information there, we say that it **writes.** In most computers, the time it takes to perform an instruction is some multiple of a basic unit called a **cycle.** Often, a cycle is the amount of time it takes to read or write one memory location.

When a CPU operates on a particular program, we say that it's **running** the program—sometimes we say that we're running the program. Since computers need to communicate with their users, computer systems include **input/output (I/O)** devices, such as keyboards, printers, TV-like screens called **cathode ray tubes (CRTs),** magnetic tape units, and speakers.

When I first began to use computers, I had trouble with the distinction between hardware and software. Some things obviously were hardware—I could touch them. But I was confused about everything else, because I didn't know how to tell for sure. My confusion was well founded: the distinction between hardware and software is arbitrary. One system's hardware is another system's software.

For example, multiplication can be performed in two ways. In some systems it's performed by hardware within the CPU—a complicated electronic circuit accepts two numbers and produces their product. But in other systems, the most the hardware can do is add two numbers. Whenever a multiplication product is needed, it's produced by a computer program—software that adds the two numbers repeatedly. The user can rarely tell the difference.

In general, the choice between hardware and software is governed more by engineering trade-offs than by fundamental differences. A typical decision is based on such factors as speed, cost, and expected frequency of use. One might guess that more elaborate operations are always performed by software. This is often true, but it isn't necessarily true. For example, many computers can play a decent game of chess, and most of these do so by means of elaborate software and no chess-oriented hardware. But some chess-playing computers have special hardware that assists in choosing chess moves.

Some appreciation for the distinctions among hardware, software, CPU, memory, and peripherals can be gained by making an analogy between a computer system and a person. It's a dangerous analogy, because the differences are more numerous and important than the similarities, but it can be useful provided you don't take it too seriously. In terms of this analogy, our brain is the CPU and memory; our eyes, ears, and hands are peripherals. When we learn new procedures and new information, we are in a sense acquiring software. The hardware is what's left for the autopsy.

Bits and Bytes and Binary Numbers

In contrast to people, who do arithmetic by using the decimal number system, most computers use the binary system. All number systems represent numbers as a sequence of digits—in different number systems, a different quantity of symbols is available for each digit. In the decimal system, ten symbols are available for each digit; we use the symbols 0–9. In the **binary** number system, only two symbols are available for each digit; we use 0 and 1. Consequently, the binary representation of a number requires more digits than the decimal representation of the same number—for example, the standard binary representation of the decimal number 32,800 requires sixteen binary digits (1000000000100000). This is cumbersome, which is why people use decimal numbers. Computers, on the other hand, can easily keep track of many digits. More important, however, is the technological convenience of the binary system—it's easy to build electronic devices that have two stable states. These states, which are analogous to the two positions of a light switch, are used to represent the two possible values of a binary digit. Those electromechanical relays in the Mark II computer—one of which beat that first bug to death—were there to represent binary numbers. For convenience, a binary digit is called a **bit**.

Numbers and Symbols

Despite their reputation, computers are not perfect calculators. We represent numbers in computers by assigning a particular bit pattern to each number, as in the standard representation of 32,800 that I mentioned before. But there are practical limitations on the number of bits that can be devoted to the results of calculations. A limited number of bits has a limited number of possible bit patterns—for example, sixteen bits provides $2^{16} = 65,536$ different patterns—but the number of possible calculation results can easily exceed the number of available patterns. Consequently, the precision of most computer calculations is limited. More bits can yield more precision, but practical limitations arise quickly. Computer systems are designed to provide sufficient accuracy for their intended applications, but accumulated inaccuracies can still cause trouble.

Many people think of computers as glorified programmable adding machines, and they don't understand how computers are used to process text rather than numbers. But if you assign the letter A to the number 1, the letter B to the number 2, and so on, you can represent any sequence of typewriter keystrokes as a sequence of numbers. In a more general sense, the numbers are irrelevant. A computer memory is a device for recording and retrieving symbols; whether those symbols represent numbers or letters is a matter of interpretation.

Measuring Storage Capacity

Because programs and data are stored in computer memories as patterns of bits, it makes sense to use bits as a measure of storage capacity. For convenience, capacity is often expressed in units of eight bits each, called a **byte.** One byte is the amount of memory that's usually used to store one character of text. Thus, for example, the word "byte" requires four bytes of storage. This entire book required about 500,000 bytes for the "raw text"—i.e., not including storage for various typesetting specifications. Another common unit of storage is a **word.** This is usually an integral number of bytes, and it usually indicates the total number of bits on which the CPU can operate at one time.

More about Memories

Because fast memory devices are more expensive than slow memory devices, the memories of most computer systems are divided into three main parts: a small amount of high-speed memory, a larger amount of medium-speed memory, and a still larger amount of slow memory. The combination of all three parts is referred to as a memory **hierarchy.**

The highest-speed memory usually is located within the CPU itself, where a few locations called **registers** are used to store frequently needed data or the intermediate results of computations. The next level of memory, called **main memory** or **primary storage,** is where the current program and data are usually stored. Typical main memory sizes are between 8,000 bytes (written 8K bytes or 8KB, and spoken as "eight kilobytes") and 10,000,000 bytes (written 10M bytes or 10MB, and spoken as "ten megabytes"). Main memory that can be both read and written is called **random access memory (RAM).** Main memory that can be read but not written is called, appropriately, **read only memory (ROM).** The contents of ROM are permanent, in the sense that they are still there after turning the computer off and then on again. The contents of RAM, on the other hand, usually disappear when the computer is turned off. ROM is used to store programs and data that are needed frequently and don't change. Because information stored in ROM has some of the characteristics of both hardware and software, it's sometimes referred to as **firmware.**

Computers usually don't have enough main memory to store all of the desired programs and data. Whatever doesn't fit in main memory is put into the third and slowest level of memory, called **secondary storage.** The information in secondary storage is usually organized into named units called **files.** A typical computer might have 100 to 1000 times as much secondary storage as main memory. For personal computers, the two most common media for secondary storage are cassette tape and floppy disks. **Cassette tape** units use standard audio cassette tapes to record bits instead of sound. **Floppy disks** look like small phonograph records encased in a stiff paper or rigid plastic jacket. They come in several sizes, from about three to eight inches in diameter. They are made of thin (floppy) material with a magnetic coating that makes it possible to record information. For reading or writing, the disk is placed in a unit that spins it underneath a magnetically sensitive device called a "head." Cassette tapes and floppy disks also serve as media for distributing programs and data—when you buy a new program for your personal computer, you're most likely to get it on a floppy disk.

Hard disks provide more capacity and higher speeds (and higher prices) than floppies; as the name suggests, **hard disks** are like floppy disks except that the magnetically coated disk is rigid. Until recently, hard disks were available only for large computer systems, but the so-called **Winchester** hard-disk technology is now available for personal computers.

The organization of information within a computer's memory hierarchy is roughly analogous to the organization of information in your study or office. As you work, you may make short side-notes on a notepad (registers). Whatever you're currently working on tends to be spread out on top of your desk (main memory), and everything else tends to be stored as files in your desk or in file cabinets (secondary storage).

Computer Terminals

Any device that combines a keyboard with a printer or CRT is referred to as a **computer terminal.** The particular combination of a keyboard and a CRT is called a **video display terminal (VDT).** A principal characteristic of a VDT is that the user gets no permanent record of the displayed information. In contrast, terminals that incorporate a printer are referred to as **hard-copy** terminals. So-called **dot-matrix** printers work by fashioning the printed characters out of a pattern of dots. Because the individual dots are usually discernible, the print quality is lower than that of a typical typewriter. Printers capable of typewriter-like quality are referred to generically as **letter quality** printers, and many of them operate like typewriters by striking raised letter forms against a ribbon. In a common variety called **daisy-wheel** printers, the letters are located at the ends of the spokes of a rapidly spinning wheel. A newer type of letter-quality printer is the **ink-jet printer,** which prints by directing a tiny spray of ink drops. Even newer (and more expensive) is the **laser printer,** which prints by a Xerox-like process after a laser has written an image of the desired text or graphics on a copying surface.

Categories of Software

Software for a typical computer system comes from several sources: some is provided by the computer's manufacturer, some is purchased from independent sources, and some is written by users. One important category of software, usually provided by the manufacturer, is the computer's **operating system;** this is the software that manages the computer system's resources. For example, the operating system interacts with the user, it coordinates the transfer of data between primary and secondary storage, it decides where in memory to store the various programs and data, it schedules the execution of programs, and it keeps programs from interfering with each other. Operating systems are sometimes called **control programs** or **executives.**

The other main category of software comprises **applications programs**—programs that perform specific functions, such as word processing, accounting, drawing, game playing, and education. Applications programs can be purchased from the computer's manufacturer or from independent sources, and they can also be written by users themselves. In most cases a user initiates the execution of an applications program by instructing the operating system to read that program in from secondary memory and start running it. When the program finishes or when the user interrupts it, control is usually transferred back to the operating system. Apart from games and education, the three most popular categories of applications software for personal computers are data base management, electronic spreadsheets, and word processing.

Data Base Management

An important and common function of computers is to maintain files of information, from telephone numbers to business inventories to social security recipients. Such computerized files are known as **data bases;** the programs that store, retrieve, and manipulate the information in data bases are known as **data base managers.** A common example of a data base is the ubiquitous personnel data base, which contains lists of employee names, addresses, salaries, and related information. The information in most data bases is grouped into aggregate units called **records;** for example, all of the information about one employee usually is contained in one record of a personnel data base.

Electronic Spreadsheets

The exploding popularity of the personal computer in businesses has many explanations. But more than anything else, the explosion was detonated by a single computer program, called VisiCalc. It is said to be the world's most popular program. VisiCalc, its various copies (VisiClones in computer jargon), and its various descendants are referred to generically as **electronic spreadsheets.**

When you use an electronic spreadsheet, you see on the screen a set of numbers arranged into rows and columns in the manner of an accounting spreadsheet, a budget, a financial balance statement, a tax return, and the like. The rows and columns have explanatory labels as headings. For example, the row labels might be "Number Ordered," "Unit Price," "Tax Rate," "Discount," "Total Price," "Labor Costs," etc., all the way down to "Net Profit." The column headings might be "1980," "1981," etc. Suppose you want to see how a proposed change in the tax rate would affect your profits in 1984. Using appropriate commands, you select the number that indicates the tax rate for 1984 and you change it. Before your eyes, other numbers on the spreadsheet change to corresponding new values whenever they depend directly or indirectly on the tax rate.

A common way of describing the advantages of electronic spread-sheets is to say that they make it easy to ask **"What if?"** questions. What if the tax rate changes? What if my labor costs increase by 10 percent? What if my orders go down 15 percent at the same time?

Word Processing

Computer systems are often augmented with special hardware and software designed to support a particular application. Word processing is an important example—indeed, it is arguably the most important application of personal computers.

A **word processor** is a computer system with features that facilitate creating, revising, formatting, and printing text. The simplest word processors essentially are typewriters with memory. When you use one to type a page of text, a copy of everything you type is saved in the memory—patterns of bits in the memory represent characters in the typed document. Later, you can revise the text by typing in changes, and the revised page can then be typed out automatically.

More elaborate word processors use a CRT to display a portion of the text currently stored in memory. This has the advantage of showing the results of changes to a document without having to print it out. Indeed, a word-processing ideal (rarely achieved) is to print documents only once, in final form. When using a word processor with a CRT, the current position in the document—i.e., the place where new text would be entered if you started typing—is shown by a little symbol that blinks on and off, called a **cursor.**

Many personal computers can be used as word processors simply by running a word-processing program. When no special hardware is involved, the word-processing program is sometimes called a **text editor.** Systems that are intended to be used primarily as word processors often include special hardware. The most common examples of such hardware are so-called **function keys.** These are extra keys or buttons, with labels like INSERT, DELETE, SEARCH, and PRINT. You ask for a particular operation by pressing the key with the corresponding label. Function keys make it easier to remember which operations are available and easier to ask for one.

Another common way of achieving these effects is to display the currently available operations as a list on the screen and to provide the user with some means of indicating which one is wanted; such lists are widely referred to as **menus.** Unlike systems that operate by means of function keys, menu-driven systems don't require special hardware. Menus are popular with all kinds of software, not just word processors. The reason for using the term "menu" is obvious, but its origins are obscure. Some say it arose from a common second love of the hacker: eating Chinese food.

Like practically anything else you can buy, word processors have a wide range of prices and capabilities. They differ in the size and visual quality of the display, the size and number of documents you can work with, the quality of printing, and the speed of operation. They also differ in terms of how well they support such basic functions as inserting text, deleting text, moving text (both within and between documents), searching for specific text, replacing occurrences of specific text, and changing the format of documents. Some word processors will merge address lists with form letters, check spelling, and perform basic data processing functions such as sorting lists or adding columns of numbers.

Another important difference among word processors is the extent to which what you see on the screen when you edit a document is a precise image of what will appear on the printed page. Some word processors have special symbols that appear on the screen to indicate the beginning or end of such special effects as underlining, boldface type, subscripts, right-justified margins, and the like. The special effects are seen when the text is printed, but the text on the screen is displayed without them. Other word processors display the text with the special effects shown on the screen exactly as they will appear on the printed page. Word processors that take the latter approach are often described by the long but relatively accurate term **"what-you-see-is-what-you-get."**

CHAPTER 4

User-Interfaces,
Friendly and Unfriendly

The history of technology abounds with devices that have become progressively easier to use. Cars, radios, and computers are all good examples. As part of the progression, various control functions are automated, less technical knowledge is required of the user, and less technical information is presented to the user. Thus, in cars the manual choke disappeared and automatic transmissions became popular. Likewise, many oil pressure and water temperature gauges have been replaced by dashboard lights. The owner's manual calls them warning lights, but we know them as idiot lights; indeed, the process of introducing such lights is known generally as "idiot proofing."

Although there is a strong economic motivation for idiot lights, they also make driving easier, and not just for idiots. Indeed, they make driving easier for thoughtful people who would rather not think about the radiator temperature. These people don't care how hot the engine is, they just want to know if it's too hot. The idiot light presents them with this information and does so effectively. Besides, most people wouldn't bother to monitor a temperature gauge—experience tells them that, unlike the gas gauge, the temperature gauge is rarely important, but they do notice when the red light comes on. Since an idiot light distracts you when the radiator temperature requires attention, you don't have to distract yourself in order to check it periodically. This allows you to concentrate on other concerns.

Some people prefer gauges. They belittle idiot lights on the ground that one can do a better job driving with gauges. This is undoubtedly true in special circumstances, but generally true only as a rationalization. For these people, the added complexity is worth it because the gauges are occasionally useful and always enjoyable. Those who prefer gauges are often experts who enjoy getting the most out of their machines; they view driving as an end rather than a means.

Computers are not so different. Like cars over the years, computers are getting easier to use, and in some respects the changes are analogous to the replacement of gauges by idiot lights. And, like drivers who favor gauges, some computer users belittle the trend toward easy-to-use systems. These users appear to thrive on complexity. They are often experts who enjoy getting the most out of their computers; they view computing as an end rather than a means.

Unlike cars, easy-to-use computers aren't called idiot-proof, they're called **user-friendly.** (As a marketing achievement, this terminology ranks with "Palmetto bugs," which is a term used in Florida— at the instigation of some genius in the real estate industry, I'm told—for large, flying cockroaches.) To an extent, user-friendly computers are indeed idiot-proof, in the sense that they reduce the amount of technical knowledge required of the user and they make it relatively hard to make certain mistakes. But there's much more to it than that, which is why I'm devoting two chapters to the subject.

Moving On from Keyboard Paralysis

If only it were enough to take a deep breath, pay attention to instructions, and start typing. But for many people, the real struggle begins when they do start typing. Having asked themselves

"Will I break it?"

"Will I destroy some information?"

"Will I feel foolish?"

"Will I look stupid?"

"What if it beeps?"

they find their worst fears confirmed. While they may not actually have broken something or destroyed information, they often think they have, which is just as effective in pumping their anxiety. First-time users often find the computer to be complicated, cryptic, quixotic, and unreliable. They type gingerly, on eggshell keycaps. It's a frustrating experience.

The Inadequacies Beyond

Pilot error is the most common cause of aircraft accidents, at least according to the results of official FAA investigations. An accident report might admit that a chart was misleading, but it blames the pilot for being misled. Another report might admit that it was easy to flip the wrong switch under the circumstances, but it blames the pilot for doing so. The situation with computers is somewhat analogous. If a user does something that damages a file or crashes a system, then it's the user's fault, even if the mistake was easy to make.

It's obvious to new users that *something* is inadequate. But what, or who? In most cases, it's not the user, although stupidity is as comfortable at the keyboard as it is at the controls of an airplane or car. More likely, the user's training is inadequate, the computer hardware or software is inadequate, or the designer of the computer system is inadequate. Unfortunately, anxiety-prone neophytes are likely to interpret these inadequacies as their own.

Computer anxiety is not the only barrier to computer facility. Indeed, the transformation from keyboard-paralyzed neophyte to full-fledged, bit-chomping hacker is in many ways a transformation from a struggle with oneself to a struggle with the system. The struggle is pervasive. From the layout of their keyboards to the ways in which they perform arithmetic, from the languages we have created to communicate with them to what we have programmed them to say, most computer systems have serious design flaws. And it is these design flaws, rather than fundamental limitations, that make computer systems difficult to use. Although novice users tend to think that only novices find computer systems difficult to use, design flaws make it difficult for users at every level of expertise. The veteran, pausing only for a sarcastic outburst, easily vaults the flaws that trip the novice; but other flaws lie in wait.

Many of the novice's problems can be solved by sufficient training, but to blame them all on inadequate training evades a basic point: Many office and personal computer systems are inadequate for the class of users that are supposed to find them useful.

Computer Personalities

Remember the distinction between Type A and Type B personalities? One was compulsive and driven, the other easygoing and relaxed. To classify people in terms of such distinctions is naïve, but it can be instructive.

I think there are two kinds of computer users; I'll call them Type 0 (zero) and Type 1. Type 0 personalities liked secret codes when they were children—many still do. They would rather solve a puzzle than read a novel, and they would rather fix or improve a tool than use it to help them with a job. Many of them were hi-fi hobbyists, and they still like temperature gauges in their cars. Type 1 personalities are different—they dislike puzzles and jargon, and they're more interested in their jobs than they are in their tools. The Type 0 personality prefers a tool with lots of options and adjustments, knowing that flexibility is useful in honing the tool to fit a wide range of jobs. For the Type 1 personality, the additional flexibility is counterproductive.

Many personal computers are no more than fancy puzzles for the Type 0 personality. There's nothing wrong with this, provided it's what you want. Indeed, personal computing is one of the most fascinating, engaging, and rewarding hobbies around. Some personal computers are also sophisticated tools that, when used capably, can help with many jobs. The Type 0 personality easily achieves this capability and enjoys doing so. The Type 1 personality finds it harder to achieve the capability, dislikes trying, and hates being forced to try.

In these oversimplified terms, it's easy to see the main problem with many office and personal computer systems. They were designed and built by Type 0 personalities, and they're suitable for use by Type 0 personalities. But they are sold to Type 1 personalities, they are forced on Type 1 personalities, and they are advertised as meeting the needs of Type 1 personalities (they don't).

The Barrier of Complexity

I love music. I played the trumpet in high school, and, years later, I studied piano for six years. I had always wanted to play Chopin. And I did, but each piece took me months to learn. I came as close to sight-reading Chopin as I have to translating Chinese into Sanskrit.

With each new piece, it was the same story: The page of notes stood squarely between me and the music. I could see the notes, I knew the music was there, but I couldn't get to it. I proceeded, haltingly, one note at a time, without playing music. And when finally I was able to play a piece passably, I could do so only in the key written. If I tried to play it in another key—say, two notes higher—it was almost like starting over. For my teacher and countless others, it's different. Hand them some difficult music, pick a key, and out flows gorgeous music. The difference is simple—their expertise enables them to reach around the notes and grasp the music. From my experiences with playing the piano, I have reached a simple conclusion: If you have to think about the notes, you can't play the music. In this respect, music and computing are somewhat alike—if you have to think about the word processor commands, you can't compose a letter.

If you approach the computer to use it as a tool, you have a goal. Just as my goal in approaching a page of notes is to play music, your goal in approaching the computer might be to analyze a budget, draw a diagram, or compose a letter. But opaque error messages, unpredictable behavior, complicated, hard-to-remember commands that seem either mysterious or irrelevant—all of these can stand between you and your goal, just as the notes stand between me and my music. This is true whether you approach the computer with enthusiasm or with fear.

Complexity is not just a problem for novices. Veterans have some advantages—error messages are more revealing to them, the computer's behavior is more predictable, and its commands are easier to remember. But an overly complex system distracts even veterans from analyzing budgets, drawing diagrams, and composing letters. In many current computer systems a barrier of complexity stands between users and their goals. To surmount that barrier, the novice has to know too much and the expert has to think too hard.

Interfaces and User-Interfaces

Unnecessary complexity in computer systems is most obvious in what's called the "user-interface." The noun "interface" has been used technically in physics and chemistry for a hundred years to refer to the common boundary, usually a surface, between two adjacent regions. For example, the boundary between the water in your sink and the air above it is known as an air-water interface. The term is useful because what goes on at the interface differs markedly from what goes on on either side.

In electrical and computer engineering, "interface" also refers to a boundary, in this case the boundary between two pieces of connected equipment. Here, an **interface** defines the communication that can take place across the connection—it includes such things as the number of wires, the voltages on each wire, the timing of each signal, and the meaning of each signal. In referring to the activities intrinsic to designing and building a working interface, engineers often use the term "interface" as a verb. Computer engineers, for example, sometimes interface a CPU and a printer so that the results of calculations can be printed. To interface a CPU and a printer is to do much more than plug them together—indeed, you can't connect them usefully until a common interface has been defined, designed, and constructed. In this context, "interface" is a useful verb. (Its bureaucratic usage, however, remains indefensible.)

The equivalent terms **user-interface** and **human-machine interface** refer to the interactions between computers and their human users. The terms refer more to the form or means of interaction than to the results of the interaction. For example, both the Apple IIe computer and the Lisa enable you to print a document stored in the computer's memory. In the case of an Apple IIe that's running the Apple Writer II word-processing software, you do the following:

a) while holding down a specially marked key, type 'P'
b) type 'NP'

In the case of the Lisa, you use a mouse to point to the word Print on a menu. (I didn't mention specific software for the Lisa, because there's no choice involved.) Both systems have a print command, but quite different user-interfaces. By almost any reasonable measure, it's easier to print a document using the Lisa. In a small way, this illustrates the important point that one's ability to accomplish useful work with a computer system depends not only on the set of possible operations, but also on the user-interface.

It's important to realize that both hardware and software contribute to the user-interface—the Apple Writer II program could have allowed or required the user to type 'PRINT' instead of 'NP', but in 1983 it couldn't have involved a mouse because the Apple IIe didn't have one. (Today, you can buy a mouse for the Apple IIe.) It's also important to realize that hardware contributes to the user-interface, not just in terms of the physical capabilities it provides, but also in terms of how comfortably it provides them. (The analysis of user-interfaces from the viewpoint of body posture, body movement, eye strain, and the like is the concern of a field called **ergonomics.**) The user-interface, however, is more than just the hardware plus the software; it's how the overall combination of hardware and software interacts with the user. Besides the obvious physical and operational components, the interaction has psychological and artistic components. This means that there are aspects of a user-interface that you can't touch, can't see, can't measure well, and can't describe well. To use a trendy term, the user-interface is—from the user's point of view—the computer's *gestalt*.

Friendship in User-Interfaces

The pedant in me doesn't like new terms. When I first heard of "user-friendly" systems, I sneered. What this commercial neologism meant, I said to myself, was merely "useful," "pleasant-to-use," and "easy-to-use." I also disliked the misleading anthropomorphism—if anything was being friendly, it was the people who designed the system and wrote the software. But I've warmed up to the term. After I caught myself saying "user-friendly" a few times, I looked up "friend" in my *Concise Oxford Dictionary*. I found terms like "sympathizer," "helper," and "patron," terms that describe well how a good user-interface should act.

What makes a user-interface friendly? Suppose you're thinking of using a particular computer system as a word processor in your office or home. In making your decision, you'll probably ask questions about what the system can do and how fast it can do it. These questions may determine whether or not the system can in principle do what you want. But you should also ask questions like:

Is it easy to remember what operations are available?

Does the system behave as described in the documentation?

Does the system include features to help me when I forget?

Is it easy to perform the operations I'll use a lot?

Is it pleasant to perform the operations I'll use a lot?

Can I predict what the system will do when I press a key or type a command?

Is the system reliable? Can I count on it to carry out the available operations correctly?

Does the system explain what's going on when there's a long delay?

Is it hard to make bad mistakes?

Can I recover from mistakes?

When the system encounters a bug or some other problem, does it try to explain what went wrong and suggest what might help?

If most of the answers are yes, then the system is user-friendly and is likely to do what you want in practice as well as in principle. If many answers are no, however, then the reduction of principle to practice is likely to involve a struggle.

The Kitchen Genie

In the pre-Cuisinart days of my youth, I kept seeing ads for an amazing kitchen device. I think it was called the Kitchen Genie. "It chops, it slices, it shreds, it grates, and much much more for only $4.72! In New York call Murray Hill-5-5300. . . ." I never called the number, but I did once get such a device. It did everything it was supposed to, after a fashion, but it was designed poorly and consequently difficult to use. If I just followed the directions, which in a sense described the Genie's user-interface, I couldn't chop with it at all. But I was able to overcome some of the Genie's obvious design flaws—by making periodic adjustments to the mechanism, to my hand position, and to my chopping motions, I could get through an onion or two. I could not, however, just chop, and the frustration wasn't worth the tears. The same thing happened when I sliced, shredded, or grated. Eventually I gave up, threw the thing out, and bought a good knife.

A poorly designed user-interface erects a barrier of complexity that hinders the user. This is true of Kitchen Genies and electronic genies, and it is true for all users, even experts. Consider the error messages I used as examples in Chapter 2:

DOS SYNTAX ERROR

FATAL ERROR . . . ILLEGAL OPCODE AT B372

MEMORY VIOLATION . . . PARITY ERROR AT A3F2

ILLEGAL INPUT . . . FATAL ERROR . . . JOB KILLED

FORMAT ERROR

SYNTAX ERROR . . . ILLEGAL TERMINAL SYMBOL

FATAL ERROR . . . REGISTER OVERFLOW AT AF45

These messages are part of the user-interface. Given the contexts of word processing and magic tricks in which they arose, the messages attest to a poorly designed user-interface.

Veterans often can see beyond the user-interface to the underlying problem—a trivial, noncomputer example is how I coped with the Kitchen Genie (to use computer jargon, I hacked around the problem). But a badly designed user-interface still distracts veterans from word processing and magic tricks. Novice users are in a much worse position. For one thing, messages such as those above alarm the anxiety-prone. But even secure novices are stopped because they don't have inner resources of expertise to summon when problems arise. For novice users, a poorly designed user-interface is more than a distraction—it can be a barrier that is either insurmountable or simply not worth climbing.

My First Byte of an Apple

As I've already mentioned, I chose the Apple Lisa as a word processor for writing this book. Like most new products in the computer field, the Lisa was late. But it was worth waiting for, not just because I ended up liking the Lisa, but also because of what I learned while waiting. To ease the wait, the computer store lent me an Apple IIe computer and a copy of the Apple Writer II word-processing program. This was a generous loan, and I welcomed it; it helped me get started, and it gave me an opportunity to see what this popular personal computer and popular word-processing program were like.

I was appalled. I knew that I had been spoiled by larger and fancier systems—I knew that I wouldn't be able to work on large documents, I knew that some operations would be relatively slow, and I knew that other operations wouldn't be available at all. But I wasn't prepared for the user-interface, which made it so difficult to use the system that I considered not bothering. With the Kitchen Genie, I couldn't just chop. With Apple Writer II, I couldn't just write.

There are two basic methods by which people revise documents stored on a word processor. One way is to revise the document directly on the word processor by exploring changes from the keyboard. The other way is to mark up a printed copy in the traditional manner and then type in the changes. To an extent the choice between these is a matter of personal preference—mine is for the first method—but whichever method you prefer, you can't use it efficiently unless the word processor has a good user-interface.

For my taste, Apple Writer II failed to support either method well. I couldn't compose and revise while sitting at the keyboard; whenever I tried, I had to struggle with the user-interface so much that I couldn't think creatively. And even when I did my creative work by marking up printouts, typing in the resulting changes was an unpleasant, laborious experience. I'll mention some particular examples in the next chapter.

Now Apple Writer II is a popular program, which implies that quite a few people find it useful, and it's true that I used it myself rather than a typewriter or pen and paper. Was my disappointment unreasonable? After all, the Apple IIe costs much less than the systems I'm used to, and the Apple Writer II software is relatively small. Was I expecting too much?

No. I was not complaining about unavoidable consequences of hardware constraints, although such constraints certainly made the designer's job harder. Rather, my complaints resulted from bad user-interface design and bad software design. Small systems with small memories *can* do a decent job on small documents.

The Knowledge Barrier

My problems with the Apple IIe and Apple Writer II were not in learning how to use them—within an hour or two of bringing them home, I was writing parts of this book. But my speed of learning was not due to a good user-interface, it was due to my previous experience. Each time I stumbled on some problem, my knowledge of other systems enabled me to proceed. When documentation was missing, unclear, or wrong, I was able to guess what was intended. If I wanted to accomplish something in particular, I knew what to look for—I knew the sort of relevant commands the system might have, and I understood computer jargon, so it was relatively easy to figure out whether I could do what I wanted, and if so, how. But I kept thinking about how hard it would be for a novice user.

Experience is a powerful resource. After using your second or third typewriter, washing machine, dishwasher, or car, the next one is easy. Poorly designed controls and bad documentation might stop the novice, but they don't stop you. The last time you rented or borrowed a car, did you read the owner's manual?

What would the first experience with a car be like for someone who had seen them around but had never even been in one? In a recent letter to one of the professional computer journals, Peter Buhr of Canada parodied complaints about user-interfaces by imagining just such an experience. Here is part of his letter:

To help me get acquainted with my car I was handed something called an owner's manual. It was filled with pages of diagrams, boxes, examples, and instructions. I tried to read it that afternoon, but got bogged down after the first two pages.

The next day the car arrived, and I jumped in for a drive. After turning it on, the owner's manual said move the gear selection lever from PARK to DRIVE. A "PARK?" A "DRIVE?" What was that? There was nothing on or near the gear shifter that said PARK or DRIVE. After spending fifteen minutes searching through the owner's manual, I realized that P stands for PARK and D stands for DRIVE. Why couldn't the owner's manual and the car agree on a labeling scheme? Why don't they say what they mean?

I drove around until evening and then came home. The next day I tried to start the car, but nothing happened. I attempted to revive the car a few times with no success. When I called the car salesman, I was told that when I turned off the ignition the headlights I was using the previous night were not turned off too. Why should the headlights remain on when the car is turned off, especially when it causes the car not to function?

A good question. Of course, "everybody" knows that you can't leave the lights on without draining the battery. Just like "everybody" knows that you can't turn most computers off without losing the contents of main memory.

Everybody, that is, except novice users. Experienced users take such knowledge for granted, so much so that it becomes a barrier. From one side, the barrier restrains and intimidates novice users. From the other side, the barrier makes it hard for experienced users to appreciate a novice user's difficulties. Indeed, novice users frequently have problems that surprise experienced users. Here's a particularly surprising example, taken from a study of problems that novice users have in learning to use word processors. One of the word processors had an operation that involved pressing a key marked EXECUTE. One participant who had had trouble performing this operation finally succeeded after several attempts and then wondered whether the initial problems had been due to hitting the EXECUTE key incorrectly. Now the idea of pressing a key incorrectly makes absolutely no sense to an experienced user. But, as the study's authors put it, "New users simply do not know enough to rule out interpretations that seem obviously wrong to an experienced user."

Despite good intentions and hard work, those who design user-interfaces and write documentation often fail because they take things for granted that need explaining. In part, these failures simply reflect old habits. For years the primary consumers of computer software and computer systems were people who already had or were willing to acquire a considerable amount of computer-related knowledge. To a large extent, experts were building systems for other experts. The Apple IIe–Apple Writer II combination exemplifies the results of such habits.

Today the balance has shifted; personal computer systems are intended more and more for use by people without much computer-related knowledge. The most successul new systems will be those with user-interfaces that meet the user on the user's side of the knowledge barrier.

CHAPTER 5

Facing the User-Interface

A good user-interface helps you to concentrate on the job you're trying to do. Flaws in a user-interface distract you; they call your attention to the user-interface and to the underlying computer system. Every computer user needs a good user-interface. But user-interface quality is especially important for the novice, whose insight about computers often stops at the user-interface. Between the novice and the effective use of computers stands a barrier of complexity and a barrier of knowledge. If you're faced with using a computer, it helps to know how good user-interfaces deal with these barriers, and how bad user-interfaces fail to do so.

Struggling at the Barriers

This is not the place for a complete review of user-interfaces. But I want to give you some examples of the kinds of flaws that are endemic.

The Obtrusive Computer

Some who use computers are tool makers, but most are tool users; they use the computer as a tool in applications that aren't inherently concerned with computers, applications like writing, accounting, and inventory control. For such applications, the computer should be as unobtrusive as possible. This means that the user-interface should fit the application well, and that it should hide those aspects of the hardware and software that have only to do with how the

86

application was programmed. Another way of saying this is that the user-interface should shield the user from the concerns of the programmer. Unfortunately, this is a difficult and rarely met goal. It hasn't been met when word-processor or spreadsheet users see messages like

FATAL ERROR . . . ILLEGAL OPCODE AT B372

or

FATAL ERROR . . . REGISTER OVERFLOW AT AF45.

Messages like these reflect abnormal conditions; they can result from mistakes by the user, from hardware failures, and from software bugs. Such problems can't be avoided entirely, but there's no point in reporting them in harsh terms that mean nothing to the user. The user may be able to solve the problem or at least work around it, but only if it's reported in understandable terms.

The underlying computer can also intrude during normal conditions. Here's an example from the Apple Writer II word-processing program: An important sequence of operations in any word processor is the one used to move text, i.e., to delete text from one place in a document and then to insert that same text elsewhere. This is the electronic equivalent of the common "cut-and-paste" editing operation performed with scissors and Scotch tape. With Apple Writer II, you can use such a sequence of cut-and-paste operations to move text, but there's an important restriction: the largest amount of text you can cut from one place and then paste in another is 1024 characters. If you want to move more than this amount, you have to go back and forth, cutting and pasting 1024 characters at a time. One thousand twenty-four characters may sound like a lot, but it's only a couple of paragraphs; I ran into the limit constantly and had to work around it.

Where did this limit come from? Well, when you cut text out of a document it gets put into a sort of holding zone, a portion of the computer's memory called a **buffer,** in technical terms. The contents of the buffer can then be pasted back into the document elsewhere. Because the buffer happens to be 1024 characters long, portions of text larger than 1024 characters have to be moved in pieces, alternately filling the buffer from one place in the document and emptying it at another.

The restriction is an annoying one—units of text comprising 1024 characters are not normally the concern of people who write books. But the existence of a buffer and the length of the buffer are normal concerns for the programmer who wrote Apple Writer II. Here's a case in which the programmer's concern protrudes through the user-interface and makes the word processor harder to use.

Perhaps you're wondering whether the protrusion is a reasonable one. After all, the Apple IIe has a limited memory, which is a fundamental constraint that must affect someone using Apple Writer II. Indeed it must. But it is the programmer's job to affect me as gently as possible so that I can concentrate on *my* job, which is writing. It is both necessary and reasonable, for example, that there be a maximum document size. But if the maximum size is ten pages and I have written only five, clearly there is room somewhere for another five pages and I should be able to move more than 1024 characters around at a time. I would have been able to do so if the programmer had used a variable-length buffer that incorporated the leftover memory space.

The Perils of Inconsistency

Not long ago I was sitting in a computer store, sampling the documentation of various popular personal computers. I happened to be sitting next to a demonstration model of one of them, and I kept noticing its keyboard—there were a variety of function keys, with labels like MENU, STORE, and PRINT. One particularly enticing key was labeled HELP, and of course I couldn't resist reaching over and pressing it. Quickly, without a delay that might have made me wonder whether I had done something wrong, the screen filled with what looked like helpful starting instructions. Here's an example (with the name of the program changed):

The best way to learn to use Goldoc is to experiment. The keys do exactly what their labels say, so don't be afraid to try something.

Great! Unfortunately, on the same screen was the following:

To protect your work from accidental loss, press [MENU] to SAVE it.

What a way to begin. I went no further.

The Apple Writer II command to cut a paragraph from a document is a simple one: you hold down the key marked CONTROL and simultaneously type 'X'. (Apple Writer II documentation uses the notation '[X]' for this, and it is known in spoken language as "Control X.") The command cuts a paragraph out of the document and puts it into the text buffer that I mentioned above. If you type the '[X]' command twice in a row, two paragraphs are successively cut from the document, but since there's only room for one in the buffer, the first paragraph is lost as soon as you type the second '[X]'. If you want to move a paragraph, you cut it out by typing '[X]', you indicate the desired position by moving the cursor to a point elsewhere in the document, and you type the command to paste (insert) text from the buffer into the document. What's the command to paste in text from the buffer? It's also '[X]'.

How can that be? How can the command to cut a paragraph out be the same as the command to paste it back in? Well, whether '[X]' means cut or paste depends on another command: if you type '[D]', the meaning of subsequent '[X]' commands change from cut to paste or from paste to cut, depending on which meaning was in effect when you typed the '[D]'. So to move that paragraph, you cut it out by typing '[X]', you indicate the desired position elsewhere in the document, you type '[D]' to change the meaning of '[X]', and then you type '[X]' to paste the paragraph into the new position. What happens if you forget to type the '[D]'? Simple: you cut paragraphs out of both positions in the document and you lose the first paragraph!

This is ridiculous. For any command to have different results in two different contexts is bad enough—it makes the word processor harder to learn and harder to use. But for a command to have a particular effect in one context and the opposite effect in another is even worse. The result, for the '[X]' command, guarantees that you will make mistakes and lose pieces of text. I lost plenty of them, and it wasn't my fault. It was the fault of the person who designed the Apple Writer II user-interface.

To print a document with Apple Writer II, you type '[P]' (Control P) and then 'NP'; I mentioned this procedure in the previous chapter. I didn't mention how to tell Apple Writer II which physical device to use when printing the document. Apple Writer II doesn't "know" this without being told because you can plug a printer into several different plugs or **slots** in the back of the Apple IIe; you have to inform Apple Writer II which slot is connected to the printer. Besides, you might have several printers connected to your Apple IIe, in which case you need to specify which one to use. To specify which printer to use, you type an additional command: if you want your document printed on the device connected to slot 4, you type 'PD4' after typing '[P]' and before typing 'NP'. For a different slot, you type the corresponding slot number in place of the "4" in 'PD4'. So far, so good. (Actually, this is already another example of a failure to shield the user from programming concerns, since a word processor user should only have to say what kind of printer to use and not which physical connection to use.)

Sometimes it is convenient not to print a document on paper, but to save an electronic image of how the printed document looks. With Apple Writer II such an electronic image can be written onto a floppy disk by typing 'PD8' before typing 'NP'. Is this because the disk drive is connected to slot 8? No! The Apple Writer II manual explains:

> **By the Way:** You type PD8 because there is no slot 8 on the Apple IIe. So when Apple Writer sees a PD8 it knows you want to print to a disk and asks you for a file name.

So, instead of typing something that directly reflects your desire to write an electronic image onto a floppy disk, you have to type 'PD8'—a cryptic command that was most likely added to the original design as an afterthought. About the only good thing to say about this command is that it's a good example of a bad user-interface.

Other perils of inconsistency occur across different computer programs rather than within a single one. For example, the user-interfaces of word processors are often inconsistent with the user-interfaces of electronic spreadsheets. Both programs include commands for printing results. But even though the commands are closely related in function, often they are not related in form. Similarly, one moves sentences or paragraphs of text in word-processing documents and columns or rows of numbers in spreadsheets, but, again, the relevant commands often are dissimilar. Such inconsistencies make the programs harder to learn and harder to use.

Rigidity Hurts

People are different. When you shop for a car, you probably look for a different user-interface than I do—different dashboard layouts, different controls, different seats. And people change. When we drive, we adjust the seat position, the steering-wheel tilt, the rearview mirror angle, the windshield wiper speed, as well as our use of such options as cruise control and radar detection. We make these adjustments in response to fatigue, to changes in road conditions, and to changes in our level of driving experience.

Similar things are true of computer user-interfaces. There is no universally satisfactory user-interface. Different users like different keyboard layouts, different types of mice, different display colors, different display formats, and different commands. And their preferences can change both within a single session and across many sessions. Flexibility is an important characteristic of user-interfaces.

Have you ever seen a picture of an antique telephone? I mean the kind that was used by holding an earphone to one ear and speaking into a microphone that was fixed to a wall-mounted box or a tabletop stand. To use it involved severe physical constraints, because you had to position your head to speak into the microphone and you couldn't move the microphone. Contrast this with the liberation of the modern handset: you can put your feet up, move around, search through your desk—all while talking.

The history of VDTs shows a similar pattern. In early VDTs the keyboard and the CRT screen were part of the same physical unit, with the keyboard firmly attached in front of and below the screen. Today VDTs increasingly are made with detachable keyboards. The keyboard comes as a separate unit that attaches to the CRT unit by means of a flexible cord much like the one that extends from your telephone handset. The IBM PCjr has taken this idea to new lengths by eliminating the cord—the keyboard communicates with the computer by means of infrared radiation. Like the modern telephone handset, the detachable keyboard is liberating. It lets you squirm and slouch, bending a hardware aspect of the user-interface in response to your changing moods and physical needs. Numerous studies have shown that the detachable keyboard is a significant help in avoiding fatigue, aches, and pains.

Flexibility is just as important in the software aspects of user-interfaces. A feature that other users like may annoy you; a feature that you like when you begin using a system may annoy you once you're familiar with the system; and a feature that you like when you use a system for one purpose may annoy you when you use the system for another purpose. Such differences are natural, and a good user-interface is designed to accommodate them.

Here's an example: The Apple Writer II command to cut a paragraph from a document is easy to initiate, but many users dislike having to memorize cryptic commands like '[X]'. They want command names that are more meaningful. An alternative user-interface could provide this by means of a single function key marked COMMAND. When you press the COMMAND key, the prompt

ENTER COMMAND:

appears on the screen and you proceed to type in the name of the command you want—in this case you might type 'DELETE PARA-GRAPH'. Given the choice, many novice users will take this alternative user-interface—in some overall sense it's easier for them to press the COMMAND key and type 'DELETE PARAGRAPH' than it is to type '[X]'. This is especially true if pressing the COMMAND key also results in a menu of possible commands. It is, however, unlikely that any user will prefer the second user-interface for long. It's too verbose.

What you welcome as a gentle introduction to a system easily becomes a burden once you're an experienced user. A good user-interface anticipates this problem by providing abbreviations and shortcuts. With this flexibility, experienced users can bypass operations that have become cumbersome rather than helpful. A common approach is to allow you to abbreviate commands by typing only the first few characters of each word—enough so that there is no ambiguity about which command you want. Thus 'DEL PAR' might suffice instead of 'DELETE PARAGRAPH'. If this were the only command to delete text, and if no other command began with the letter "D," than a simple 'D' would suffice.

Another example of shortcuts occurs in the user-interfaces of the Lisa and the Macintosh, both of which rely heavily on the inherent flexibility of mice and menus. Users can initiate almost all commands by pointing to a menu's name with the mouse and then pointing to a particular item on the corresponding menu (the menu "pops" up when you point to its name). This procedure was designed carefully, and it's always fast. It can nevertheless become tedious, especially in the case of frequently performed commands like cutting and pasting. So shortcuts are provided for such commands—you can initiate them from the keyboard by holding down a special key while you type a single character, just like the '[X]' command in Apple Writer II. To help you in remembering these shortcuts, they are listed on the corresponding menus.

A different approach is to allow you to assign a name to any particular sequence of commands. You can then use the assigned name in place of the so-named sequence of commands. The assigned name serves as a user-defined abbreviation. For example, suppose that your system has the commands 'Close', 'Print', and 'File' respectively for finishing editing, for printing, and for storing a document. If you notice that you often invoke these three commands in sequence, you might assign a special name to the sequence—say, 'CPF'—so that you could type the command 'CPF' instead of 'Close' followed by 'Print' followed by 'File'. If you think of the assigned name as a new command, what you have is a rudimentary form of programming—the ability to define new commands in terms of old commands. Some systems allow the user to do this with menu selections as well as typed commands. It's an extremely flexible and powerful feature.

Reading verbose messages from the computer can be just as annoying as having to type in verbose commands. At first, verbose messages can provide welcome, detailed information about what the system is doing and what went wrong. When directed to the novice user, these messages can be user-friendly. But once you've seen them a few times, they become unnecessary, boring, and time-consuming. At this point you want to see the kind of unfriendly, terse messages that you didn't want to see in the beginning. Some user-interfaces anticipate this problem by starting with verbose output, and allowing you to request terse output later. A too-helpful computer can get in the way, just like a too-helpful person.

Lowering the Barriers

We know of 1,250,000 species of animals, 260,000 species of vegetables, and 3,000 types of minerals. These facts are small examples of the enormous complexity inherent in zoology, botany, and mineralogy. Yet despite their complexity, these fields are routinely studied, expanded, and taught because we have discovered a structure that makes their complexity manageable. The difference between chaos and order is often a difference in manner of description or point of view.

Complex human activities such as writing, financial planning, and accounting are likewise manageable if you exploit a suitable structure. And structure is the key to avoiding complexity in the user-interfaces of computer programs that support these activities. Complexity in user-interfaces can arise from two sources: the human activity being supported and the computer system being used to support it. A good user-interface brings order both to the job being done and to the computer that is helping to do it.

The Magical Number Seven

A key to managing complexity in human activities is recognizing the role and the limitations of human short-term memory. Most people cannot deal competently with a great many things at the same time. Whether we're in our office or our kitchen, if there're too many things going on at once we get "frazzled"; we spend too much time keeping track of what's going on and not enough time dealing with it. The evidence for this "frazzle factor" is widespread: managers in traditional organizations rarely have more than five to ten people reporting directly to them. It's easy to remember seven-digit phone numbers and five-digit zip codes, but it's hard to remember twelve-digit credit card numbers. And even with phone numbers, it's harder to remember 2029361472 than 202-936-1472. Structure helps by partitioning possibilities into manageable chunks.

How many chunks of information are manageable? What makes a chunk manageable? How do people organize information into manageable chunks? These are interesting and important questions. They are also difficult questions—our knowledge about their answers hasn't changed much in the twenty-five years since George Miller published his fascinating and now famous paper, "The Magical Number Seven, Plus or Minus Two: Some Limits on Our Capacity for Processing Information." Miller summarizes a variety of experiments that measure our ability to process information. His conclusions: we can receive, process, and remember only about seven things at a time. Although we exploit various methods in overcoming this limitation, there is a real "informational bottleneck" associated with the number seven, a number that also has an intriguing presence in the days of the week, the notes of the musical scale, the wonders of the world, the ages of man, the deadly sins, and the levels of Dante's hell.

Making the User-Interface Manageable

The first step in keeping the complexity of a user-interface manageable is to prevent the underlying computer from being a source of unnecessary complexity. The computer should be unobtrusive. Second, the user-interface should not require you to learn everything before you start. You should be able to do useful work without knowing all of the commands and without knowing every possible articulation of the commands that you do use. The full power of the user-interface should be disclosed progressively as your needs and interests develop. As the American computer scientist Alan Kay put it:

Simple things should be simple; complex things should be possible.

A third characteristic of a manageable user-interface is that it limits the current alternatives and helps you to remember what they are. One obvious way to do this is to keep the interface simple—if there are only seven possible operations, for example, most people will be able to remember them, provided that these possibilities aren't multiplied by inconsistencies and exceptions, and provided that their effects don't depend on previous actions that have long since faded from short-term memory. Visual reinforcement can help greatly in reducing the load on the short-term memory. Common examples are the use of screen menus and function keys. Another general approach is to provide **help facilities**—on-line documentation that appears on the screen when you ask for help, perhaps by pressing a HELP key or by typing in a special command. On-line documentation can be particularly helpful if it is tailored to the situation that exists when you ask for help.

Literally a new dimension in visual reinforcement has become available with the development of high-resolution displays. Typical computer terminal CRTs used to be **character-oriented,** able to display a fixed set of characters in a fixed set of positions—for example, popular character-oriented CRTs display 24 lines of 80 characters with 96 choices available for each character. Recently, systems like the Star, the Lisa, and the Macintosh have been developed with displays that allow the CPU to turn on and off individual dots on the screen; because each dot corresponds to a bit in a special portion of memory, such displays are called **bit-mapped** displays. Thus, instead of a display containing 24 lines of 80 characters each, the Lisa's display contains 364 lines of 720 dots each. These dots can be used to fashion characters, but they can also be used for arbitrary graphics.

In the Star, the Lisa, and the Macintosh, bit-mapped displays are used to display small, suggestive pictures called **icons.** They are used much like the imaginative graphics that appear in some car or appliance manuals. For example, different icons distinguish word-processing documents from spreadsheet and other documents. In conjunction with a mouse or other pointing device, icons can serve as **screen buttons**—sort of electronic function keys. To use the Lisa or Macintosh as a calculator, for example, you begin by pointing with the mouse to an icon that looks like a small electronic calculator and you press the button on the mouse.

A fourth and related way in which user-interfaces can be kept manageable is by making it easy to perform frequent operations. For example, Apple Writer II and many other programs use simple control-key combinations like those I mentioned in earlier examples. Function keys and screen buttons also serve this purpose, as do various means for quickly selecting an operation from a displayed menu. High-resolution displays are useful here in yet another way: instead of typing the name of an operation explicitly—or implicitly by means of a function key, screen button, or menu selection—you can initiate some commands by moving objects around on the screen. For example, with the Star, you can print a document by using the mouse to move the document's icon over to a printer icon.

Dealing with User-Created Complexity

Complexity arises not just from a computer system's capabilities, but also from the results of applying them. People create large numbers of different kinds of documents, and these can be difficult to manage, whether they're in a file drawer or in a computer memory. A good user-interface will help you to manage the complexity you yourself create. Many of the features that provide flexibility in a user-interface can be exploited to manage user-created complexity.

An important capability for managing user-created complexity is provided by systems that allow you to group related documents together, give the group a name, and then manipulate the entire group in terms of that name. This is analogous to placing related paper documents into a labeled file folder. Taking this analogy a little further, some systems provide even more structure by allowing you to put file folders inside other file folders—such systems are said to provide **hierarchical** or **tree-structured** files.

Some systems permit the user to divide the CRT screen into separate regions that act somewhat like independent terminals. A common application is to provide simultaneous views of several files or simultaneous views of different parts of the same file, which is why the regions are known as **windows.** By relieving the load on the user's short-term memory, windows are useful in managing user-created complexity. High-resolution, bit-mapped displays with windows and icons are particularly useful; they exploit the well-known human ability to perceive visual patterns quickly, analyze them quickly, and remember them. In many respects it's easier to deal with a two-dimensional arrangement of objects on the screen than with a list of the same objects, just as it's easier to deal with large arrangements of papers on top of your desk than with large files inside your desk.

Dealing with the Knowledge Barrier

The most obvious and effective way to deal with the knowledge barrier is to keep the computer unobtrusive. Whenever technical details about the computer hardware and software can be seen through the user-interface, it's hard for users to function effectively without understanding those details. Yet to hide the computer completely usually is neither possible nor advisable. The more users know about computers, the easier time they will have. And the more knowledge about computers the designer can assume users to have, the easier time the designer will have. The trick is to find the right balance, which depends on the hardware that's used, the kind of person who is the intended user, and the complexity of the application. But whatever level of computer knowledge the user is assumed to have, it's important to maintain that level consistently throughout the user-interface.

Experience is a powerful resource. Computer veterans deal easily with user-interfaces because their experience provides conceptual models that help them to learn quickly, answer questions, reduce uncertainty, and predict behavior. These advantages aren't normally available to novices because their experience hasn't provided them with the same models. In an attempt to retain these same advantages for novices, some user-interfaces are designed explicitly in terms of models that are based firmly on noncomputer experiences. The Star and Lisa user-interfaces, for example, deliberately model a typical office environment. There are electronic counterparts to familiar kinds of documents as well as to familiar operations on documents, such as creating, editing, copying, printing, filing, and mailing.

The Design of User-Interfaces

You're an experienced programmer. Two years ago you were hired by WordSoft, a new company specializing in word-processing software for personal computers, and your first job there was to develop a word-processing program for the new Admiral-64 computer system. Before you started writing and testing the program, you were required to get approval for your design, and your boss gave you two months to prepare for the "design review."

Typically, you spent those two months in two interrelated activities: first, choosing and specifying the set of operations that will be available to users of the program; and, second, designing the internal structure of the computer program. Your results passed the design review with distinction, but they didn't lead to a widely acclaimed program. Why not? Because you left out another activity, just as important as the other two: specifying how the program will interact with the user—i.e., designing the user-interface.

The importance of user-interface design is recognized widely, but such recognition is relatively recent. Although there are notable exceptions, hardly anyone used to pay much attention to user-interfaces. One reason is that, in the past, most computer users were experts, if not hackers. Expert users tend to view user-friendly interfaces with disdain rather than relief. For them, user-friendly interfaces are sometimes limiting and never essential.

Another reason for the lack of past attention to user-interface design is that the predominant style of computing used to be different. For years, most computing was **batch**-oriented—users submitted their jobs to the computer, went away, and came back later for the results. Today the predominant style is **interactive**—users sit in front of a computer terminal while they work. Their inputs cause various computer actions, and they react to the resulting outputs. With batch computing, the user and the computer work separately on the user's job; with interactive computing, they work together. User-interfaces are correspondingly less important for batch work than for interactive work. A user-interface can't interfere as much with your thinking if you do your thinking away from the computer.

Even after the predominant computing style became interactive, most computer terminals were until recently "line-oriented" as opposed to "screen-oriented." With a **line-oriented** terminal, the interaction between user and computer consists of the repeated exchange of one or more lines of text—commands from the user to the computer and responses from the computer to the user. Typewriter-like computer terminals are all line-oriented. Many VDTs, despite their TV-like screens, are also line-oriented—in computer slang they're called **glass teletypes,** a sarcastic reference to the limited capabilities of the venerable Western Union devices that served for years as computer terminals. With a **screen-oriented** terminal, the interaction between user and computer encompasses the entire screen, with all the associated advantages of two dimensions over one. Screen-oriented terminals enable user-interface improvements that are impossible with line-oriented terminals. The huge electronic-spreadsheet market shows how valuable such improvements can be.

One other reason for the lack of attention to user-interfaces is that improving a user-interface often expands the size of the software and makes it run slower. When the basic resources of a computer system—memory and computer time—were much more expensive than they are today, it seemed wasteful to consume them in the user-interface, especially since the high cost of computers usually guaranteed that computer experts would be the primary users.

Today things are different. Users interact primarily by means of screen-oriented terminals, and computing resources are cheaper. More to the point, there are more computers than computer experts, and most computers cost less than computer experts. Consequently, there is an overwhelming economic incentive to make the computer useful to a wide variety of users. For this a good user-interface is crucial, a fact that had led to the growing demand for and investments in good user-interfaces—build a better mouse-trap, and the world will beat a path to your interface. Yet many designers still don't pay systematic, disciplined attention to the user-interface. They think carefully about *what* things the system will do and *how* they will be done, but not so carefully about *for whom* they will be done. A common result is that first-time users find the computer to be complicated, cryptic, quixotic, and unreliable.

Providing good user-interfaces is the only way to expand drastically the universe of computer users. In the past, commercial pressures centered on making computers do more; today they center on making computers more accessible. This change is recent; the term "user-friendly" is only a few years old. But the change has occurred quickly and permanently. Yesterday's exception—the computer without a computer expert—is proving to be today's rule.

Easier Said Than Done

To design a good, user-friendly interface is not a casual undertaking, and it can't be done as an afterthought. Computers like the Star, the Lisa, and the Macintosh have excellent user-interfaces. But Xerox spent more than thirty person-years of effort on the user-interface alone, and they designed the user-interface before the Star hardware was built. Moreover, they worked for two years before writing any of the final product software. The experience at Apple was similar, although they were able to take advantage of Xerox's efforts.

Some of the difficulties arise from the task that the software is intended to support. The more complicated the task, the harder it is to design a good user-interface. There are, for example, word processors with good user-interfaces for routine writing tasks. We have several at the Naval Research Laboratory, and we use them a lot. But we don't use them to write papers containing complicated equations, complicated tables of data, and large numbers of bibliographic references. Their user-interface just isn't sophisticated enough to support this kind of writing. Instead, we use a large computer system and five computer programs. We have to use five different notations to describe what we want to see in the printed paper, and we engage in a frustrating amount of trial and error in getting the results to look just right. We're about as far away from the "what-you-see-is-what-you-get" style of word processing as you can get. The user-interface is lousy, but the overall process is much better than typewriting.

Software designers have many goals: They strive to provide numerous and elaborate capabilities; they strive to make the software correct, fast, and easily modifiable; and they strive to make the software easy to use. Unfortunately, these goals are highly interrelated, and this fact is responsible for much of the difficulty in user-interface design. For example, given whatever price they're willing to pay, people want the fastest, most capable, and easiest-to-use systems. But speed costs. Like high-performance cars, sailboats, and airplanes, the fastest software is often uncomfortable and limited in overall capability. Furthermore, capability and ease-of-use are in part available only at each other's expense, and the definitions of both qualities involve personal preferences.

If you get a word processor with lots of features, you may have trouble remembering how to use them. Unless you're a keyboard virtuoso with instant recall and lots of computer experience, beware. As the number of features grow, so do their interactions, and in the end the growth can be counterproductive. Adding features to the word processor paradoxically can make it less capable—it might become slower, more cumbersome, and harder to manage. It doesn't take much to transform elegant simplicity into baroque complexity that makes the system harder to use and more error-prone. With a word processor that has more "features," you may actually get less work done. It all depends on the user-interface.

From another point of view, the problem with user-interface design is that there are few scientific or psychological principles on which to rest firmly. It's one thing to say that the user-interface should be fast, powerful, simple, elegant, consistent, helpful, error-resistant, and flexible; but it's another to measure these qualities and to understand how they interact. Whether or not something appears to be consistent, for example, often depends on your background and point of view. The date February 29th is either a glaring inconsistency that occurs once every four years, or it is the key to consistency between our calendar and the earth's position relative to the sun.

In practice, the design of a good user-interface occurs largely by trial and error. Although intuition about the desired qualities can be a helpful guide, the best way to evaluate design proposals is to test them. Yet such user-testing is difficult, lengthy, and not always conclusive. Both Xerox and Apple spent a lot of time testing different mice. Xerox settled on a two-button mouse, while Apple settled on a one-button mouse. At Apple, even the names of the Lisa commands were determined in part by user testing. The design of user-interfaces is as much an artistic as it is a scientific activity.

In contrast to the old adage about books and covers, a user-interface reveals a lot about the design of the underlying software. Indeed, the difficulties of software design compound the difficulties of user-interface design. Flaws in the software structure are hard to hide behind the user-interface. Inconsistencies in the user-interface usually reflect corresponding inconsistencies in the structure of the underlying software; this is bad because it increases the chances of bugs and it makes the software harder to modify when improvements are wanted. The Lisa, for example, keeps track of the last time you modified each document. But when the Lisa first came out, merely looking at a word-processing document counted as a modification, whereas spreadsheet documents weren't considered to be modified unless you actually changed them. Such an inconsistency burdens the user, who has to remember two different definitions of "modified," and it makes the software harder to change. The criteria for a concept like "modify" should be expressed in only one place within the software—in the Lisa there were at least two such places.

The best way to smooth out a user-interface that has such inconsistencies is to clean up the software structure. If you try to smooth out the user-interface without removing basic flaws in the underlying software, the flaws will still be felt by the user, like a pea under layers of royal mattresses.

Some Advice for the Novice User

All computer users struggle with the consequences of design flaws, whether the flaws are deep within the hardware and software, or at the user-interface. In particular, a bad user-interface is difficult to learn, difficult to use, and error-prone; it interferes with useful work. These problems affect all users, but they are most significant for the novice user.

Advice for Those Who Can Choose

If you're in a position to choose hardware, software, or both, concentrate on the user-interface. Make sure that you get the capacity and the functions to meet your needs, but concentrate on how effectively you'll be able to use that capacity and those functions. If a particular feature is important to you, don't just ask *whether* it's included, ask to see *how* it's included. In evaluating your choices, insist on a thorough demonstration and some hands-on experience. Prepare yourself in advance with a stock set of tasks, then see how the system can be used to accomplish them. (This is a little like making the rounds of stereo equipment stores with your favorite record under your arm.) Do the same with the documentation. Prepare yourself with a stock set of questions, then see how long it takes you to find the answers. Remember that there's usually a trade-off between capability and ease-of-use—if you're willing and able to put up with a more difficult (less friendly) user-interface, it may be advantageous to do so. But the trade-off is a personal one. Remember also that your feelings and your needs will change. Try to anticipate these changes by choosing a system that has appropriate flexibility.

Like other choices that we make, choosing a computer system would be much easier if we could go back later and choose again. Unfortunately, few of us can afford to do so, and money-back guarantees aren't in vogue. But you can afford to talk with those who preceded you, and this is one of the best ways to get reliable information. Seek out others like yourself. Listen to what they say about the system they're using. Ask them about the user-interface.

If you conclude that there aren't any user-friendly systems that meet your needs, or that you can't afford the ones that do, then wait. User-interfaces are one of the fastest-changing aspects of the fast-changing computer business. Costs will continue to come down, and user-interfaces will continue to improve.

Advice for Those Who Can't Choose

Even if you can't select the user-interfaces, understanding their role and knowing about typical flaws can help you to deal with them. Use this information, for example, to control your anxieties and frustrations. It's easier to deal with problems when you know that they're not your fault. If you have to use a system that's too hard for you, ask for help. You deserve it.

You can also use your general knowledge about user-interfaces to make the best of a bad one. Design your own "crib sheets" and paste them onto your terminal. Make sure that you understand all of the flexibility that's available, and exploit this flexibility to mold the interface in ways that are useful to you. A useful feature in this regard is the ability to create your own commands by naming a sequence of existing commands. Find patterns of usage that are the least error-prone and that help you in managing complexity. Seek out others who may have found comfortable patterns. Read the documentation carefully. You might uncover something that will make a difference. Learn some more about computer hardware and software. Ask for special training. Lower the knowledge barrier.

Bad User-Interfaces Are Not a Fact of Life

Many people appear prepared to accept poorly designed user-interfaces because they think "that's how computers are" or "that's computer language." But bad user-interfaces are not the result of fundamental limitations and properties—they're the result of bad design and badly written software. Unfortunately, people are encouraged to believe otherwise by many personal-computing enthusiasts, computer salespeople, and computer magazines. Consider the following excerpt from "Some People *Should* Be Afraid of Computers," an article that appeared in the magazine *Personal Computing:*

> A loud beep and a terse message can be unnerving, but these are the only ways the computer can point out an error. The messages are vague, but that's a matter of necessity.

Nonsense. Necessity has nothing to do with it. Terse, vague messages are the result of design choices. The author goes on to explain:

> With a limited amount of memory space available for storing error messages, each message must be designed to cover a wide variety of error situations. Since you can't see your actual mistake, it's natural to want a specific description, but not getting one is a fact of life with personal computers.

Not getting specific descriptions of your mistakes is neither a fact of life nor a fact of electronics. It is true that personal computers have relatively small memories, and this fact does force the software designer to make difficult choices. The designer of a word-processing program, for example, must decide how much of that memory to use for the word-processing instructions, how much to use for error messages, and how much to use for the report, letter, or poem the user will write. But if the error messages unnerve you and are also too vague to be useful, you haven't encountered a fact of life—you've encountered a system that isn't for you. For you, it would have been better had the designer chosen to reduce the size of the largest possible document in order to include more detailed error messages. Better yet, the designer could have exploited the knowledge that it takes brand-new users a while to start writing large documents. When a small document is being written, the leftover memory could be used for better error messages. The software could then provide detailed error messages until you're ready to trade them in, so to speak, for the ability to write larger documents.

The computer can be the most useful tool you've ever used. Realizing its potential, however, is not automatic. It depends on the kind of partnership you form with the computer. And more than anything else, it's the user-interface that determines this partnership. With the right user-interface, you can waltz. With the wrong one, you'll wrestle.

CHAPTER 6

Conversing with Computers

As an indicator of intellect, language can be misleading. Many small children pick up sophisticated phrases and use them in exactly the right context without the slightest idea of what they mean; their felicity can be confused with wisdom. Many foreign visitors speak halting English; their infelicity can be confused with stupidity.

So it is with computers. Many computer users encounter facile communications and they misjudge the underlying level of understanding. They are quick to attribute intellectual capabilities to the computer that it doesn't have and can't have. Other users find it enormously difficult to get a computer to do their bidding—communicating is a constant struggle, which they lose as often as they win. They conclude that the computer's reputation as an effective tool is undeserved. "Computers are supposed to be so capable, but this one acts so dumb!" These two extreme reactions are both inappropriate for today's computers. The reactions result from expectations formed in conversations with people, expectations that make it harder to cope with computer anxiety and harder to use computers effectively.

When you use a computer to analyze a budget or write a report, you're engaged in a conversation of sorts, a conversation that takes place across a user-interface. But the language of that conversation is extremely limited, and the limitations can be a constant source of struggle and frustration. The limitations reflect technical and economic constraints, such as the speed of the CPU, the speed of the memory, the size of the memory, and the amount of effort devoted to user-interface design and software development. The limitations also reflect a lack of knowledge—regardless of other

108

constraints, we don't know *how* to program computers to interact in all of the ways that might be convenient and effective.

If you understand these limitations, you're better able to deal with them. Understanding can also head off some of the frustrations, especially those caused by expectations that carry over from conversations among people. Here is another case where anthropomorphic attitudes cause problems.

Eliza Does Little

A computer program called ELIZA is perhaps the most famous example of misleadingly facile communication between people and computers. ELIZA was written in 1966 by the American computer scientist Joseph Weizenbaum. Weizenbaum was studying the use of natural languages like English in user-interfaces, and he wrote ELIZA to show that a relatively simple procedure can give the appearance of understanding. ELIZA mimics Rogerian psychoanalysts—those who make their couch-resident patients do most of the talking. ELIZA can be pretty convincing, especially if you're receptive, and it quickly became popular. It's still popular. Versions of ELIZA and ELIZA-like programs abound (DOCTOR is another common name).

Naturally, when I was drafting these paragraphs I couldn't resist generating an example. So I had a few "conversations" with a version of ELIZA. The first three were pretty dumb—few would find them convincing, I hope—but I started falling for the fourth one. Here is the transcript, exactly as it appeared on my terminal screen. The lines in the transcript alternate between my inputs and the program's responses. The responses have no capitalization and no terminal punctuation marks, but the omissions aren't too distracting. Besides, I prefer to show it to you as I saw it. The transcript begins with my first input to the program:

My book is getting to be a major problem.

please continue, this may be interesting

It's all because of my problem with procrastination.

have you any other problems you wish to discuss

Well, of course I want people to like the book.

i am not sure i understand you fully

Well, I would just die if people think it's a stupid book.

do you think it's likely that people think it's a stupid book

I don't really know. But if they do, it would be agonizing.

do you wish that they do

Of course not! Do you think I'm a masochist?

we were discussing you—not me

Oh yeah . . . I'm even more worried that I won't even finish.

please go on

It's the procrastination—I can't seem to control it.

what does that suggest to you

Perhaps I don't want to finish the book.

And so on.

ELIZA is a relatively simple program, without any pretensions of understanding either input or output, yet it can participate in some surprisingly cogent exchanges. It works by recognizing certain key words in the input sentence—words like "problem," "if," and "you"— performing some simple transformations on the sentence, and printing out the result or some stock sentence like "i am not sure i understand you fully." The program has some ability to remember special words or phrases that appear in the input. If it gets stuck later, this stored information enables it to stimulate further conversation with something like "tell me more about your mother."

ELIZA showed clearly that one can mimic intelligent responses without understanding the ongoing conversation. This wasn't really a surprise; as I mentioned already, we know that children can do it. Grown-ups do it too—for example, to cover up during inattentive telephone conversations. Yet it was interesting to see that the behavior could be captured in a computer program—often, writing a computer program serves to verify that we understand what has been automated. Besides, some of ELIZA's more effective conversations were surprisingly long.

The real surprise, however, was in people's reactions to ELIZA. People became emotionally involved—Weizenbaum's secretary, engrossed in a conversation with ELIZA and feeling the need for privacy, once asked him to leave the room. Some people believed that ELIZA demonstrated a general solution to the problem of understanding English with a computer. And some psychologists even believed that the program could evolve into automated psychotherapy. These reactions to style without substance and form without content remind me of the Peter Sellers movie *Being There*, in which society exalts the cryptic utterances of a well-dressed, elegant moron. The worst part about the reactions to ELIZA is that they occurred despite Weizenbaum's repeated explanations of the program's purpose and methods.

One can speculate at length about people's reactions to programs like ELIZA. Plausible explanations could be based on the psychology of conversation, on the urge to trust machines, and on the urge to anthropomorphize computers. Perhaps ELIZA attracts people by the technological aura of machine-dispensed wisdom. Whatever the reasons, ELIZA exemplifies the kind of facile communication that can mislead the novice user.

Do What I Mean, Not What I Say

Unlike children, dandelions, and cockroaches, computers do exactly what they're told. This may sound like a good thing, especially to those who are about to use computers for the first time. But a bit of experience, with computers or even with children, is enough to demonstrate otherwise. Having a child do exactly what you say is not the same thing as having a child do exactly what you want. Most of us have experienced impish children who respond to instructions by choosing the most literal interpretation. You tell them to jump into bed and they do, nearly breaking the springs. You ask them whether they're happy or sad, and they say "yes." The first few times a child does this sort of thing, it's cute. Thereafter, it's annoying and frustrating. With a computer it's almost the same, the difference being that it's not cute even the first time.

It is ironic but true that the computer's strict obedience to instructions is a constant source of frustration; it contributes to anxiety in the novice and anger in the veteran. The main reason is that, while you may think you're instructing the computer properly, it's easy to get confused and issue either "illegal" instructions or legal instructions that do something other than what you really want. In short, your instructions don't always correspond to your intentions, but the computer goes right ahead and follows your instructions exactly. The result? Perhaps a cryptic error message, perhaps damaged information, perhaps a system crash.

Such problems plague all users, but novices are especially susceptible. They're ill-equipped to avoid trouble, and ill-equipped to get out of it. And when they encounter trouble, they tend to blame themselves; they interpret the designer's failures, the programmer's failures, and the computer's failures as their own. Veterans know more, and they can use that knowledge to avoid trouble and get out of trouble. Moreover, when they do encounter trouble, their egos are less likely to be damaged than their files. They remain secure because they understand how various failures can lead to trouble, and because they often understand what happened in a particular case. To the veteran, the trouble was predictable, at least in retrospect. To the novice, however, the predictable appears capricious.

Computers follow instructions, whether or not the instructions correspond to intentions. This fact is summarized by the following ditty, said to have appeared on the bulletin board in a room full of computer terminals:

I really hate this damn machine,
I wish that they would sell it.
It never does quite what I want,
But only what I tell it.

Courting Disaster and Winning

A large number of discrepancies between intentions and results are caused by confusion on the part of the user. Here's an example: I sometimes use a text-editing program that interprets what you type depending on which of two modes it's in at the time. When the program is in "insert mode," almost everything you type is interpreted as text to be inserted into your document. When the program is in "command mode," almost everything you type is interpreted as a sequence of commands for manipulating the text that is already in the document. My trouble is that I often forget that the program is in command mode, and I start typing some text I want added to my document. The program interprets every character as a command, with predictable but definitely unwanted results. Suppose, for example, that I move the cursor to a position between two particular words and type in 'deep', thinking that the word "deep" will be inserted into my document. Instead, the 'd' (in command mode) means to start deleting text; the 'e' means to do so up to the end of the current word; the second 'e' means to move to the end of the next word; and the 'p' means to put back in the last piece of text that was deleted. The result is to exchange the position of two words. Instead of inserting the word "deep," I reverse the order in which two words appear in my document.

It can get much worse. There is a story, probably apocryphal, told about another text-editing program that has separate input and command modes. According to this story, a hapless user wanted to type the word "edit" into his document. Unfortunately, the program was in command mode when he started typing—the 'e' selected everything currently in the document; the 'd' deleted everything that was selected; the 'i' caused the program to enter insert mode; and the 't' inserted the letter "t". The result: the entire document was replaced by the letter "t". Sorry.

The story may be apocryphal, but it isn't misleading. Computer systems abound with easy ways to cause catastrophic results. A famous example concerns a popular computer operating system (control program) called **unix**. Like any complete operating system, unix provides a way to delete files (without it, you would eventually run out of space). The command is called "rm" (for remove); to delete a file called "bookone," for example, you would type in the command 'rm bookone'; to delete the two files "bookone" and "booktwo," you would type in the command 'rm bookone booktwo'; and so on (the file names are separated from each other by one or more spaces). If many files are involved, this can get tedious, so you're likely to take a shortcut and make use of unix's ability to accept a "wild card"—a special character, in this case the asterisk (*), which stands for any possible sequence of characters. Thus, the command 'rm book*' deletes all of the files that have names beginning with the four characters "book." Unfortunately, people sometimes mis-type commands like this and include an inadvertent space, for example 'rm book *'. This too is a meaningful command—too meaningful. Unix responds first by deleting the file "book" (if it exists) and then by deleting all of the files whose names match the wild card, i.e., every single file!

This really happens. Unix is the operating system on one of the computers I use at the Naval Research Laboratory. The system manager told me recently that at least five people had come to him for help during the past year after inadvertently deleting all of their files in this way. Because he regularly makes backup copies of every file in the system, he had been able to restore many of the lost files. But some work was usually lost, and the experience was always annoying. Novice users find such experiences to be not just annoying, but unnerving.

If all your files are deleted because you type an extra space, then it's your own fault. But does the punishment fit the crime? The situation is somewhat analogous to that in a military aircraft cockpit where the bomb-release switch is placed, without a safety catch, next to the landing-gear switch. Similarly, if you forget the current mode of that text editor and scramble or delete existing text instead of inserting new text, that's your fault, too. But the mistake is a likely one, and you should be fair to share the blame with whoever designed the program to use separate input and command modes.

In fairness, let me mention that the designer of the text-editing program was responding to a built-in constraint—the lack of special function keys on a standard keyboard. (Indeed, the disadvantages of modes are among the reasons why many word processors use keyboards with extra function keys or augment the keyboard with screen menus and a pointing device like a mouse.) The designers of unix were likewise responding to constraints—they were striving for a small, simple but powerful operating system that doesn't protect users as much as it allows users to protect themselves. They succeeded, and the widespread popularity of unix is well-deserved.

In both cases, however, the user-interfaces make it easy to damage or destroy information. There are compensating advantages for veteran users. But the disadvantages can overwhelm the novice user, who is more likely to make mistakes and less likely to understand that the programs are just doing what they're told.

Obeying the Bugs

It was the eve of my mother's birthday, and we were on our way to a dinner party that my father was holding in her honor. Our destination was a restaurant in New Jersey to which we had never been before. My friend Susan and I were surprise guests, so it was important that we arrive before everyone else. We weren't worried about finding the restaurant, however, because we possessed typewritten instructions sufficient for a cretin leaving from the North Pole. You wouldn't believe how detailed these instructions were unless you either saw them or knew my father, who had written them. Reports on my father as a youth vary, but through some combination of predisposition, Czechoslovakian law school, and the British Army, he became a meticulous planner (psychologists have less flattering terms). When he writes down instructions, they tend to be complete, precise, and unambiguous. We make fun of his instructions, but we follow them.

On this particular night the unexpected happened: we turned right onto Route 9W, as instructed; we proceeded to the second light, as instructed; and we started to turn left, as instructed, onto what was supposed to be East Clinton Avenue. Only it wasn't East Clinton Avenue.

We did what anyone would do under the circumstances: we considered what the likely errors might be (Dad had either counted the lights wrong or gotten the street name wrong, more likely the first); we considered the obvious alternatives to getting back on track (turn left anyway and see if the instructions made sense without the street being East Clinton Avenue, or continue on 9W and see if East Clinton Avenue turned up soon); and we started trying these alternatives. In particular, we continued on 9W and found East Clinton Avenue at the next light, where we turned left. From that point on the instructions made sense again. My mother was pleasantly surprised, and we all had a good time.

Had we behaved like a typical computer, however, we probably wouldn't have found the restaurant. We might have executed the left turn as instructed and continued to interpret the instructions as best we could regardless of where they took us until we either crashed, ran out of gas, or came to the end of the instructions at the wrong building. Or we might have stopped immediately at the offending intersection and put a white handkerchief on the car. Or we might have returned to Washington, D.C.—our starting place— and tried again. Had we been in an unfriendly mood, we would have driven to the nearest phone booth, called my father, screamed

JOB ABORTED . . . FATAL REDUNDANCY CHECK!

and hung up.

Like my father's instructions for finding the restaurant, computer programs don't always express what we want them to. The resulting behavior may be surprising, but it's predictable from the programs; like the examples in which there were discrepancies between users' intentions and their instructions, the computer's behavior is exactly in accordance with its instructions. Even when an error occurs because the computer hardware malfunctions, the resulting behavior is often just as predictable.

Unfortunately, although the result may be predictable, it can seem capricious. Moreover, the cause can be hard to see. Suppose that you try to print a document, perhaps with some printing options you haven't used before, and the system crashes, taking your document with it. Whom do you blame first? Chances are it's not your fault, but it takes self-confidence to question the validity of a slick user's manual. It may also be hard to pin down the blame. If you take the time to re-create the problem and try to understand what happened, you may fail because it's easy to overlook some of the circumstances that were involved. Something you or someone else did ten minutes ago may be relevant, but you don't realize it.

We got to the restaurant because we were able to detect and handle the bug in my father's instructions. Unfortunately, computer systems are not as adept as people at detecting, handling, and reporting the problems that arise from erroneous or misinterpreted instructions. There are really two issues here: first, detecting that something is wrong; and second, doing something about it. Error detection is accomplished in almost all cases by consistency checks that exploit redundant information. My father's instructions, for example, were redundant. Instead of telling us just to turn left at the second light on Route 9W, he added the name of the street and it was this redundancy that made us notice the error. The same approach is used in computer systems to detect both software bugs and hardware malfunctions.

Redundancy is used, for example, to detect malfunctions of computer memories. In a typical malfunction, some of the memory locations can become unreliable—when the bit patterns are read from these locations, they may be different from the patterns that were written there. In effect, a number stored in the faulty location may change to a different number, or an instruction stored there may change to a different instruction. Analogous problems could have arisen with my father's instructions had I misread them or spilled coffee on them. Such memory errors are often detected using extra bits that store redundant information. In some computer systems, every 8-bit byte of the memory is accompanied by one more bit, called a **parity bit,** which is set to one or zero depending on whether the number of ones in the byte is odd or even. Whenever a byte is read from memory, so is its parity bit. If the byte is inconsistent with its parity bit, one of the two has changed, most likely the byte, and the computer system's software is so informed. Typically, the program then running is stopped and its user gets a message like

MEMORY VIOLATION . . . PARITY ERROR AT A3F2.

Parity bits help to detect memory failures but not program bugs. Other hardware consistency checks, however, can pick up bugs. One example is the phenomenon known as "overflow." Most CPUs operate on a fixed number of bits at a time—usually 8, 16, or 32 bits—and this puts a practical limit on the magnitude of the largest number that the CPU can handle in routine arithmetic operations. If an instruction attempts to generate a number that exceeds this maximum value—for example, by dividing any number by zero, or by multiplying two numbers that are both close to the maximum— the CPU register containing the result is said to **overflow,** and this fact is reported to the computer system's software. The user might get a message like

FATAL ERROR . . . REGISTER OVERFLOW AT AF45
712 547 234 232
777 234 342 455
209 487 439 332
>

More often than not, overflows result from bugs or invalid usage rather than from hardware errors. In the magic trick example from Chapter 2, the overflow occurred when a '0' (zero) was entered in response to the prompt

PLEASE ENTER A NUMBER BETWEEN 1 AND 10:

This is an example of an "illegal input"—there's no law against entering zero, but the programmer didn't anticipate the possibility and it got the program into trouble.

Bugs can also be detected by consistency checks that are included in the software itself, like the consistency checks in my father's instructions. Unfortunately, such consistency checks consume extra resources—both memory and execution time—and most programmers prefer to spend these resources on additional capabilities. Besides, although every programmer knows how hard it is to write correct programs—something I'll talk more about soon—most programmers believe that their own programs are correct. As a result, in much of the software that's in widespread use, software consistency checks are sparsely distributed and fairly general in nature. Although many program bugs eventually result in the violation of a hardware or software consistency check, considerable damage may occur first. Furthermore, the connection between a failed consistency check and the actual bug is often so remote as to be unhelpful.

Once the computer detects an error, something must be done about it. One approach is to restart the offending program, or even the entire system, in the hope that the problem won't reappear—either because it was a transient hardware error, or because it was a software bug that surfaced through a series of unlikely coincidences. But the most common approach is to have the computer system stop executing the offending program and report the error to the user.

When a word-processing or spreadsheet program stops abruptly with messages that blame parity errors, register overflows, subscript checking, protection violations, bus errors, illegal instructions, and the like, the user-interface is shattered—the messages have nothing to do with writing a letter or analyzing a budget. Such messages are certainly helpful to the programmer who wants to fix the word-processing or spreadsheet program. They are even helpful to the computer-savvy user who can exploit them in working around the problem. To the novice user, however, they are discouraging.

What Would a Person Do?

When asked to set fire to the logs in your fireplace, a friend will oblige cheerfully. Asked to set fire to your house, the friend will at least say, "Are you quite sure?" Is this kind of behavior too much to ask of a computer? Why can't a computer react to 'rm book *' more cautiously than to 'rm book*'? In fact, it can. A computer can notice that you've asked for all your files to be deleted, and it can react by asking for confirmation, but only if it's programmed to do so.

Fortunately, in most cases it's relatively easy to have a computer program detect and double-check on commands that would delete all of your files, and many current systems behave in this way. Indeed, systems tend more and more to protect users from electronic damage caused by their mistakes. Many text-editing and word-processing programs double-check with you before deleting the entire contents of your document, remind you to save copies of changes you make, and the like. Many of them have a general "undo" command that's able to reverse the effects of most single commands. It's easy to see how this might be possible—conceptually, the program could maintain two working copies of your document, one being the current version and the other being the version just prior to your most recent command. The effect of the "undo" command is to revert from the current version to the prior version.

It's also possible for computers to do a better job handling the consequences of software bugs and hardware failures. Instead of screaming

JOB ABORTED . . . FATAL REDUNDANCY CHECK!

they can be made to behave more like we did on our way to my mother's surprise party. Again, suitable programming is required. Typically, it's missing. To an extent, its absence reflects a lack of knowledge; we don't know how to write programs that handle errors as well as people do. But the absence of effective error-handling software also reflects a lack of effort; designers typically choose to spend the available CPU cycles and memory capacity elsewhere.

What about the small, repairable, but frustrating damage I did by typing 'deep' when the text-editing program was in command mode? The resulting exchange of two words isn't the kind of major change that merits confirmation, and the "undo" command wouldn't work because the computer interpreted 'deep' as a sequence of four commands. From the program's point of view, I typed in a sequence of four correct commands, so it seems hard to argue that "the computer should know better" or even that it could know better.

But a person might well suspect that I was confused about the mode. For one thing, the sequence of commands that exchanges two words happens also to spell a common English word. This can even happen for an intentional sequence of commands, but it's enough of an unlikely coincidence to raise suspicions. Having become suspicious, a person might look at my document in the vicinity of the cursor position, realize immediately that the word "deep" makes sense in that context, and conclude that I could be confused about the mode. Given this evidence, it would be reasonable to check with me before making the change.

Can a computer do this? Partially. Checking a dictionary to decide if a string of characters composes a common English word is an easy programming task. But it's another matter to write a program that can decide if it makes sense to insert that word at a given point in a document. Fair accuracy might be attained just by determining whether the word is the right part of speech required by the context—adjective, noun, verb—and people in the field known as computational linguistics know how to write programs that could do this. For example, because the word "deep" can serve as an adjective, such a program could analyze a sentence like

"He took a breath, fast becoming faint,"

and conclude that it's reasonable to insert "deep" before "breath." But the same program would conclude that it's reasonable to insert "deep" into the sentence

"The condor baby fell out of the nest,"

right before "condor." You know that such an insertion makes little sense, whereas reversing the positions of "condor" and "baby" makes a lot of sense. For that matter, reversing "breath," and "fast" in the previous sentence makes as much sense as inserting "deep" before "breath"!

These examples show that high accuracy would require programs that can analyze not only grammar, but meaning. This is extremely difficult; there do not exist programs that can analyze the meaning of unrestricted English with anything close to the sophistication of a native speaker.

We Speak Such Different Languages

Every language is a means of expression consisting of a vocabulary and a way of using it. These characteristics are true of all languages, whether natural languages that have evolved to support communication among ourselves or artificial languages that we have created for communicating with our machines. The importance of natural language arises from life; the importance of artificial language arises from technology. Mathematics is an artificial language with ancient roots, but it is not as rigorously defined as is commonly supposed; many aspects of mathematical notation are not formally defined, and a correct interpretation of the notation relies on the reader's common sense and general understanding. These informalities do not apply to the artificial languages that we use to communicate with computers. Indeed, the development of computer technology elevated formally defined artificial languages from a theoretical tool to a practical necessity, and it stimulated important advances in their theory and applications. Artificial language permeates the information age.

Syntax and Semantics

It has proved useful to describe languages in terms of two different properties: the form of correct expressions in the language and the meaning of those expressions. The technical terms for these properties are "syntax" and "semantics." **Syntax** is a set of rules for forming correct expressions. English syntax, for example, includes the rules for forming grammatically correct English sentences. An example is the familiar rule stating that simple English sentences can be formed by a noun followed by a verb followed by an object, a rule that leads to sentences like

"John drinks coffee."

The syntax of a language determines whether an expression has a correct form, but it is the **semantics** that assigns meaning to the expression. After you read that John drinks coffee, you know that a person named John sometimes pours a liquid into his mouth and swallows; you know that the liquid is made from the roasted and ground seeds of a particular plant that is grown primarily in South and Central America; you know that the liquid is probably hot and probably dark; and you know that John probably ingests caffeine when he drinks the liquid. All of these conclusions illustrate the semantics of one simple English sentence.

The concepts of syntax and semantics apply not only to the natural languages we use when we converse with each other, but also to the artificial languages we've created for conversing with computers. Here I don't just mean artificial languages we use in writing software. We also use artificial languages when we type in commands to a word processor or a spreadsheet program, and we use artificial languages when we read their responses. When we tell unix intentionally to 'rm book*' or unintentionally to 'rm book *', we are using an artificial language. And when we type numbers into the magic trick program, we also use an artificial language, in this case a simple one containing just enough syntax and semantics for expressing numbers. In fact, the definition of numbers from the Divine Calc manual (Figure 8) is an informal description of their syntax.

Syntax Errors

Computers can communicate in terms of these artificial languages because people can write computer software that determines the meaning of language expressions. Ideally, software to do this for a particular artificial language is based on a complete description of the language's syntax and semantics. When you type in an expression using the artificial language, a program attempts to parse the expression, thereby dividing it into its various syntactic components—analogous to finding the noun, verb, and object of a simple English sentence. If you happen to enter an expression that doesn't fit any of the language's syntactic rules—a common mistake known technically as a **syntax error**—the program should detect that such an error has occurred and reject the offending expression. How gracefully it does so depends on the program. I gave an example of a syntax error as part of the magic trick example in Chapter 2 when I suggested that you might respond to the prompt

PLEASE ENTER A NUMBER BETWEEN 1 AND 10:

by typing in 'SEVEN' or 'VII'. In order of decreasing grace, the various likely responses were

ILLEGAL NUMBER, TRY AGAIN:

or

FORMAT ERROR

or

SYNTAX ERROR . . . ILLEGAL TERMINAL SYMBOL

or

WHAT?

or even a shrill beep without any message at all.

Semantics—Intentional and Otherwise

Once the software has determined that an expression is syntactically correct, it proceeds to apply the corresponding semantics. In the case of the magic trick, the software proceeds to perform the trick with whatever number you entered. In the case of 'rm book*' or 'rm book *', it proceeds to remove the specified files. It's obvious that problems will arise if there are bugs in the programs that are supposed to take the actions deemed by the semantics.

Such bugs can arise from programming mistakes as well as from misunderstandings about the intended semantics.

There is, however, a more subtle source of problems: not all syntactically correct expressions are semantically meaningful. This can happen with natural languages; for example, reversing the subject and the object in the coffee example yields

"Coffee drinks John,"

which is syntactically correct but semantically meaningless. Similar problems can arise in artificial languages, and if they occur without being detected, they can result in unexpected behavior.

To use a common example, most computer systems have a command that will display the contents of a file on your terminal screen, the assumption being that the file is filled with human-readable text. To view such a file—for example, a file called "book"—you use a program that I'll call "see"; typically you type in something like 'see book'. But not all files are filled with human-readable text. Some files contain bit patterns that represent CPU instructions and other internal control codes, and these bit patterns cannot be interpreted as English characters. For example, if you happen to type 'see magictrick', where "magictrick" is the name of a file containing the CPU instructions of the magic-trick program, strange things are likely to happen. The bit patterns in magictrick are meaningful when interpreted by the CPU as instructions, but not when interpreted by your terminal as English characters. But many of the bit patterns for the CPU instructions in magictrick happen to coincide with bit patterns that are normally used to control your terminal, so when you send all of these bits to your terminal by typing 'see magictrick', your terminal tries to obey but proceeds to do wild and crazy things. Typically, random characters splatter all over the screen while the terminal beeps and flashes. When it's all over, you may not be able to resume normal conversation with the computer without turning your terminal off and on again.

The 'see book' and 'see magictrick' commands are both syntactically acceptable—to use a natural-language analogy, they both have verb-object forms. But only the first is semantically meaningful. Many computer systems nevertheless go right ahead and process the second command as though it were semantically meaningful, the result being behavior quite unrelated to the normal semantics of 'see'. This is a bit like taking the sentence "Coffee drinks John" and trying to interpret it literally, despite its being semantically meaningless.

Another example from the magic-trick program is the consequences of your responding to the prompt

PLEASE PICK A NUMBER BETWEEN 1 and 10:

by entering '0' (zero). In contrast to its treatment of 'SEVEN' and 'VII', which the magic-trick program rejected on syntactic grounds, the program accepted the '0'. It shouldn't have, as the prompt itself suggests, but it did. The program then processed the zero according to semantics inherent in the magic trick that the program was intended to demonstrate, semantics that didn't apply to zero. It did so till the bitter end, which in this case arrived with a hardware consistency check:

```
FATAL ERROR . . . REGISTER OVERFLOW AT AF45
712  547  234  232
777  234  342  455
209  487  439  332
>
```

This outcome and the outcome of 'see magictrick' are not what either you or the programmer really want. In both cases you are asking for something that's not included in the intended semantics, and in this sense the outcomes are your fault. But the programmer can prevent such inappropriate processing and, for not doing so, deserves some of the blame. Here again, the computer is doing exactly what it's instructed to do.

Artificial vs. Natural Languages

The artificial languages that we've programmed computers to understand are much smaller, much simpler, and much less expressive than the natural languages that people understand. These contrasts arise in part from difficulties in writing the computer programs that process artificial languages. These programs make progress, like other programs, by executing long sequences of CPU instructions. Step by step, the programs have to scan the input language expression, divide it into relevant pieces, determine which syntax rules apply (if any), and take whatever actions are implied by the language semantics. All of this requires that the language be defined completely by the designer and well understood by the programmer.

As the syntax becomes more complicated and the semantics more sophisticated, it becomes harder for the designer to define the language and harder for the programmer to write language-processing programs that work properly. The programs also become bigger and slower—the bigger, more complicated, and more sophisticated the language, the more instructions are required to process language expressions. And this requires more memory and either more processing time or a faster computer.

In the case of a typical personal computer system, say an Apple IIe or an IBM PC with between 64K and 512K of main memory, the artificial languages are limited severely by the computational resources available. (Keep in mind that the available resources are needed for more than language analysis.) Conversations with such systems are characterized by rigid, unforgiving syntax and simple semantics. On much larger systems, however, people have written programs that can actually converse in English, albeit severely restricted subsets of English. Moreover, these programs converse by exploiting both syntax and semantics, in contrast to programs like ELIZA, which operate almost entirely at a syntactic level. Unfortunately, the language subsets are too small to be really useful, and the computer systems are too big to be commonly available— tens to hundreds of times bigger and faster than typical personal computers. Even so, the fastest programs take many seconds to process even relatively simple sentences (which is considerably slower than human conversational speed), and they can take much longer.

In between these two extremes are programs that provide a somewhat flexible, forgiving syntax together with limited semantics. These programs are advertised as being able to converse in English, but it's more appropriate to describe their capabilities as

English-like. Nevertheless, they can provide a much better user-interface than what is typical today. They are starting to become available for personal computer systems—typically as a user-interface for data base management programs—and you will be seeing more of them.

Tomorrow's Conversations

The quality of our conversations with computers will improve. Intellectual advances will lead to better user-interface designs, better artificial languages, better methods for processing useful subsets of natural languages, and better methods for handling errors. Technological advances will make greater computational resources available at lower cost. The capabilities of today's large, institutional computers will be available in tomorrow's desktop, personal computers, so that we will be able to exploit the fruits of our intellectual advances.

One thing will not change. Your computer may make it easier for you to instruct it in accordance with your intentions, but it won't divine those intentions. The computer is a machine whose ability to communicate with its users results from basic principles of operation and from the manner in which it is programmed. Computers not only do exactly what they're told—barring a malfunction, they don't do anything without being told. Your computer won't read your mind. Neither will it love you or honor you. It will, however, obey you, for better or worse.

CHAPTER 7

The Sachertorte Algorithm

"Software" has made it into the general vocabulary. You read about it in *Time*, *Newsweek*, and *The Wall Street Journal*; you hear about it at cocktail parties; and you see it in TV advertisements. Lots of people who know almost nothing about computers know about software. They know that it's produced by people called computer programmers and that it somehow contains the computer's "instructions"—it's the stuff that transforms a computer into a video game, a word processor, or an electronic spreadsheet.

Beyond these glimpses, however, the picture is often murky. For many people, software blends into the computer's high-tech, forbidding image, which obscures the answers to such basic questions as

"What does software look like?"

"How do you write it?"

"What do computers actually do with it?"

Answers for the professional programmer are necessarily long and detailed. But less-detailed and still-useful answers are relatively easy to come by—you can understand a lot about software with just a little bit of effort. This is possible because many of the underlying concepts are neither difficult nor totally unfamiliar—we deal with them routinely in the context of many human activities, including such mundane ones as cooking, cleaning, and driving. Illuminating these concepts and relating them to computer software can help to make computers less mysterious and easier to use effectively.

The Problem of Aunt Martl's Sachertorte

Leaving home to begin adult life can be traumatic for children and parents alike, and every family seems to develop rites of passage that ease the transition. My own departure was marked by parental offerings. From my father I got the "Getting Ahead Professionally" installment of the "Facts of Life" lecture series. From my mother I received two documents, each potentially useful and each illustrating one of my mother's many qualities—the first, that my mother never gives up (instructions on how to clean my apartment), and the second, that she'll always worry about her children's comfort (a collection of family dessert recipes).

I ignored the documents for some time. When I went back to them, I of course did so for the recipes. I craved a particular chocolate cake, and baking it without my mother's help was the only path to satisfaction that seemed both feasible and respectable. When I found the recipe, however, I wasn't so sure. Here it is:

Aunt Martl's Sachertorte

5 ozs. butter, 2 ½ ozs. chocolate, 3 egg yolks,
5 ozs. sugar, 2 tablesp. bread crumbs,
½ teasp. baking powder, 3 egg whites.
Bake slowly in not too hot oven, cake should be moist.

Not much of a recipe. Such "recipes" sufficed for my mother and other experienced cooks, but not for me. So I phoned my mother, vaguely aware that this may have been her intention all along, and asked for an algorithm.

I didn't actually say "algorithm"—I asked for more detailed instructions. But algorithm is what I had in mind. And since the concept of algorithms is as basic to computer science as it is convenient in discussing my trials with Aunt Martl's Sachertorte, I'll tell the rest of this story in terms of algorithms.

Algorithms Are Everywhere

First, I need to explain what an algorithm is. It hasn't yet reached the same vogue-word status as "software", but it's also widely heard—an **algorithm** is a precise description of a method for solving a particular problem using operations or actions from a well-understood repertoire. Algorithms are everywhere. When we change the tires on a car, mow the lawn, vacuum a rug, or follow directions to a restaurant, we use an algorithm. A person can solve problems by means of algorithms expressed in English, provided the person understands English. And a computer can solve problems by means of algorithms expressed in a programming language, provided that the computer "understands" that language.

Now back to the Sachertorte. To me, the recipe was little more than a list of ingredients—a lot was missing. What about my mother? Had she memorized a more detailed recipe for Aunt Martl's Sachertorte, a recipe for which this list of ingredients served as a reminder? No. In fact, given the list of ingredients for practically any cake, my mother knew how to prepare the ingredients and bake the cake. If you think of my mother as a cooking machine (forgive me, Mother), it's clear that two of the operations in the machine's repertoire are preparing ingredients for a cake and baking the cake. In terms of these operations, the machine's algorithm is a simple one: first, "prepare the ingredients", and second, "bake the cake". This algorithm served my mother well, but it was useless for me because it was expressed in terms of operations that I didn't know how to perform myself. I needed to refine each of its steps until they were expressed in terms of operations in my personal repertoire, like "butter the pan". In consultation with my mother, that's what I proceeded to do.

In describing the process further, however, I want to express the resulting algorithms in a slightly more formal notation. A common and useful method of doing so is to lean heavily on the notation of a particular programming language. I'll use a "Pascal-like" notation, which simply means that, where possible, I'll use the formal notation of a programming language called Pascal, together with a variety of informal additions to allow me to express myself clearly. I need these informal additions in order to express commands like "bake the cake", which are not part of Pascal or any other programming language. I enclose such commands in double quotation marks, and I use the sans-serif type face to emphasize that the algorithms are intended as inputs that will instruct a cooking machine. Because algorithms expressed in this way look a lot like computer programs, and because **code** is a synonym for program (from the historical term "order-codes"), it's common to say that the algorithms are written in **pseudo-code.** In any case, here is a Pascal-like version of my mother's algorithm—I'll call it Algorithm AMT-1 (for Aunt Martl's Torte):

begin "prepare ingredients"; "bake cake" **end**

The meaning of this algorithm doesn't depend on the vertical and horizontal arrangement of statements. The layout is generally chosen to aid readability. For example, Algorithm AMT-1 can also be written as follows:

begin
    "prepare ingredients";
    "bake cake"

end

The words shown in boldface type are special symbols called language "keywords." **Keywords** have special meanings—in technical terms, they delimit syntactical units of the language and they are associated with particular semantic interpretations. In Pascal the keyword pair begin-end delimits a **compound statement**—a sequence of individual statements that are separated by semicolons. The semicolon is also a special symbol in Pascal—not just a punctuation symbol, a semicolon between two statements indicates that the first statement is to be executed before the second, where "first" and "second" refer to the standard left-to-right order of reading, as in English. Because semicolons have semantic implications in Pascal, their placement is important—in reading Pascal programs, you have to pay more attention to semicolons than you do when reading English. The expressions in quotes are my informal extensions to Pascal—they delimit informal statements that are not recognizable as being within the Pascal programming language. As long as you and I recognize them, however, they can be useful in expressing algorithms.

If You've Baked One Torte, You've Baked Them All

When I asked my mother to give me detailed instructions for "prepare ingredients", she explained that Aunt Martl's Sachertorte was a kind of torte, and that all tortes are prepared in essentially the same way. When pressed, she explained that most tortes were cakes made without flour, and that you always separate the eggs when making a torte properly, with the beaten egg whites being folded in just before baking. Moreover, there are two kinds of tortes: those that include butter and those that don't. If there's butter, you cream the butter and sugar first and then add the egg yolks. If there's no butter, you beat the egg yolks and then add the sugar. Different tortes have different flavorings (e.g., melted chocolate or almond extract) and different dry ingredients (e.g., breadcrumbs or ground nuts), but whatever they are you add them after dealing with the egg yolks and sugar, and before folding in the egg whites. Here is the algorithm, as I understood it, for "prepare ingredients" in the case of tortes—I'll call it Algorithm AMT-2:

```
begin
        "separate eggs";
        if "torte includes butter"
            then "cream butter, sugar, and egg yolks"
            else
                begin
                    "beat egg yolks";
                    "beat in the sugar gradually"
                end;
        "beat in flavorings and dry ingredients gradually";
        "beat egg whites";
        "fold in egg whites"
end;
```

As before, the layout isn't significant—I chose it to make the algorithm readable. Here I've introduced a new kind of statement, the **if-then-else.** It has the following meaning: The expression after the **if** keyword is a condition that must be either true or false; when it's true, the statement after the **then** keyword is executed; otherwise, the statement after the **else** keyword is executed. Either of the two alternative statements can be a compound statement; in the example above, the statement following **else** is a compound statement. Statements like **if-then-else** are called **conditional statements** because their effects depend on some condition—in this case, the effect depends on whether the condition "torte includes butter" is true or false. Conditional statements are a key ingredient in any programming language.

I referred to the **if-then-else** in Algorithm AMT-2 as a "statement" and yet it clearly contains other statements. This is potentially confusing, but only if you don't keep in mind which unit is being discussed. If we're conducting a telephone fund-raising campaign and I ask you for the "second number" on a list of five telephone numbers, I expect to hear the seven digits of the second telephone number and not the second digit on the first telephone number. Similarly, Algorithm AMT-2 is a compound statement that contains a sequence of five statements. The **if-then-else**, which itself contains several other statements, is the second of these five statements.

Algorithm AMT-2 bordered on being useful to me, but I was still unsure about how to accomplish certain operations, one example being "cream butter, sugar, and egg yolks". To find out how, I decided to stop bothering my mother and start consulting *The Joy of Cooking.* I found an informative description; it included some advice about temperature, and it suggested that the eggs be beaten in one at a time. Here is *The Joy of Cooking*'s algorithm for creaming butter, sugar, and egg yolks, which I'll call Algorithm C (for Creaming):

```
begin
      "bring butter to about 70° F.";
      if "room temperature is very hot"
            then "put the bowl in a deep pan of 60° F. water";
      "beat butter in bowl until smooth";
      "beat in the sugar gradually";
      repeat "add egg yolk to bowl"; "beat well"
            until "no more egg yolks";
end
```

This algorithm includes an example of an **if-then-else** statement without the **else** clause. Leaving it off like this is a short hand way of saying

```
if "room temperature is very hot"
      then "put the bowl in a deep pan of 60° F. water"
      else "skip to the next statement";
```

Also, to express my intention that the eggs be beaten in one at a time, I've used a Pascal feature that facilitates the repeated execution of one or more statements, namely the **repeat** statement. In any **repeat** statement, there is a sequence of statements between the keywords **repeat** and **until,** often called the **body** of the **repeat** statement; the body is executed repeatedly until the expression immediately following the **until** becomes true—this stopping condition is checked after each repetition. The **repeat** statement is an example of a general class of statements called, for obvious reasons, **repetitive statements.**

The Culinary Subroutine

I started with Algorithm AMT-1, my mother's cryptic but effective solution to any cake-baking problem,

```
begin
      "prepare ingredients";
      "bake cake"
end,
```

and I refined it twice. First, I expanded "prepare ingredients" into a sequence of five statements (Algorithm AMT-2). One of these statements was "cream butter, sugar, and egg yolks", which I further expanded into a sequence of five statements (Algorithm C). The overall result can be expressed by substituting Algorithm C into Algorithm AMT-2, yielding Algorithm AMT-3:

```
begin
        "separate eggs";
        if "torte includes butter"
            then
                begin
                    "bring butter to about 70° F.";
                    if "room temperature is very hot"
                            then "put the bowl in a deep pan of 60° F. water";
                    "beat butter in bowl until smooth";
                    "beat in the sugar gradually";
                    repeat "add egg yolk to bowl"; "beat well"
                            until "no more egg yolks";
                end
            else
                begin
                    "beat egg yolks";
                    "beat in the sugar gradually"
                end;
        "beat in flavorings and dry ingredients gradually";
        "beat egg whites";
        "fold in egg whites";
        "bake cake"
end;
```

This represents considerable progress. Although "bake cake" remains unexpanded, the rest of the algorithm is relatively detailed and useful.

This same approach could be taken with my mother's other recipes, but at least one aspect of the results would prove to be unpleasant. Creaming butter, sugar, and egg yolks is a common operation in dessert recipes. If every recipe that required you to "cream butter, sugar, and egg yolks" also included the detailed instructions for doing so—the instructions corresponding to the compound statement in the **then** clause in Algorithm AMT-3—the recipes would be tiresome. It would be better to keep these instructions separate as Algorithm C, and merely refer to them when needed by saying "cream butter, sugar, and eggs", as in Algorithm AMT-2.

Indeed, many cookbooks use brief instructions like "separate eggs" or "whip cream" in their recipes, while they expand these instructions with detailed explanations presented elsewhere. This approach has several advantages. To begin with, it saves space. Confining the detailed explanations to special sections also makes it easier to find and learn the meaning of "separate eggs" or "whip cream". Furthermore, the cookbook is easier to change. If a new method of creaming butter comes along—e.g., with a food processor—it's easy to incorporate the alternative into a revised edition of the cookbook, because the change need be made in only one place. Also, if several methods of creaming butter are described in the special section, a recipe written with "cream butter" makes it clear that the choice is up to the cook. Finally, the recipes themselves are easier to read because they are uncluttered with details that aren't needed most of the time.

I have just described, by culinary analogy, one of the oldest and most important programming concepts—the subroutine. A **subroutine** is a program whose effects can be obtained in other programs by referring to the subroutine by name whenever its effects are needed. Usually, the text of the subroutine is separate from the texts of the programs that make use of it. A program that uses such a subroutine is said to **invoke** it or **call** it. For example, Algorithm C is a subroutine called by the saying "cream butter, sugar, and egg yolks" in Algorithm AMT-2. Subroutines are also called **subprograms** and **procedures.** Using subroutines can help to keep the text of a program reasonably short and uncluttered, thereby making the program easier to read and understand. Because only one copy of a subroutine is needed no matter how many times it's invoked, subroutines save memory space. They can also make software easier to change. If a better algorithm is found for an effect that is confined to a subroutine, one can switch to the new algorithm by changing the subroutine and without changing all of the programs that use the subroutine.

A Truly Refined Recipe

I could go on from Algorithm AMT-3, continuing to refine the meaning of statements like "**beat egg whites**" and "**bake cake**" until every statement comprised instructions within my personal repertoire, but I've already illustrated the basic idea. This approach to solving the Sachertorte problem illustrates a well-known method of programming known as **stepwise refinement** or **stepwise program construction.** Generally speaking, you start with a simple, even trivial algorithm for the program's intended effect. You then refine the algorithm repeatedly until all of the informal statements are gone and every step is expressed as a statement in the repertoire of the programming language being used. Because one can think of the initial algorithm as being at the top of a descending series of increasingly detailed steps or levels, this approach is also known as **top-down programming.**

Once an algorithm is expressed completely as a Pascal program, it can be executed on computers that "understand" Pascal, i.e., on computers that have been programmed to interpret correctly the meaning of Pascal programs. In fact, the same Pascal program can produce the same results on different computers, even if the computers have CPUs that can't execute the same instructions. To see how this happens requires information about what computers actually do with computer software, which I'll explain soon. But first I want to make sure that I don't leave you with the impression that everything about programming is easy. I don't want to lead you astray, so I'll shift direction slightly.

Some First Hints of Danger

Having looked at a few, I can tell you that there's a great variety of popular books about dieting. They differ in gimmicks, they differ in style, and they differ in visual appearance. There are diet books for protein lovers and diet books for grapefruit lovers. There are pedantic diet books and frivolous diet books. They come in different sizes and with a remarkable variety in their use of lists, tables, diagrams, etchings, and photographs. Amid all this variety, however, there is one common message: Dieting is easy. You can do it.

While I was thinking about writing this book, prudence drove me to Washington's bookstores, where I browsed through the sections devoted to computer books. It was a depressing experience, and not only because there was an astonishing number of computer books. The variety of gimmicks, styles, and visual forms reminded me at once of diet books, a depressing enough subject itself, but I was most depressed by another similarity—the prevalence of a common message: Programming is easy. You can do it.

Actually, I have mixed feelings about this message. To some extent, these books are not misleading their readers. Many of the basic concepts involved in programming are easy to understand, and there's plenty of evidence that it's easy to learn how to write simple programs. Millions of people have learned how to do it, many of them from that plethora of *Programming Made Easy* books. Indeed, part of *my* message is that the basic concepts of programming are easily understood. Nevertheless, there's a lot more to programming than the stepwise refinement of Aunt Martl's Sachertorte might suggest.

One misleading aspect of the Sachertorte algorithm is that I expressed it in terms of instructions from the repertoire of cooks (even unsophisticated cooks), while these instructions are not in the repertoire of computers (even sophisticated computers). When a cook follows a recipe, he constantly exploits experience, intuition, and prior knowledge to make up for what is in fact an incomplete description of a solution to the cooking problem at hand. Moreover, most cooks are unaware of the extent to which they do so. In this regard, the Sachertorte algorithm is misleading because it glosses over two crucial aspects of programming—understanding in full detail how to solve a particular problem with an algorithm, and expressing that solution completely and precisely. To appreciate how difficult these can be, you need only try to program a real computer to cook. In part to dramatize this point, the American computer scientist Brian Reid started with a recipe from Julia Child's *Mastering the Art of French Cooking* and attempted to write a program that a gourmet robot could use to cook beef Wellington. Reid—no slouch programmer—gave up after spending twelve hours producing a sixty-page program that dealt with only one fifth of the recipe. An experienced cook is comfortable with such instructions as "slice the meat", "season to taste", "spread evenly", "cook until tender". But try to express them in sufficient detail to instruct a mindless robotic servant—an electronic Amelia Bedelia—and you'll give up too.

The Sachertorte algorithm is misleading in other ways as well. For example, I left out all sorts of details that require attention when writing a real program, I barely touched on ease-of-change, and I ignored efficiency. Moreover, I ignored the most important and difficult aspect of programming: program correctness.

A common problem with rudimentary programming examples and experiences is that they can mislead people about the difficulty of writing correct programs. It's easy and fun to write a small program that appears to work correctly. It's considerably harder to be sure that the program is correct. But writing a large program and being sure that it's correct is extremely hard. In Part III I will talk more about the general problem of software reliability, but here I want to show you two small examples that hint at the dangers.

A Subtle Difference Can Lead to Trouble

Pascal has other repetitive statements besides **repeat**. For example, the same effect as

```
repeat "add egg yolk to bowl"; "beat well"
until "no more egg yolks";
```

can be achieved with a **while** statement:

```
while "eggs yolks are left" do
   begin
     "add egg yolk to bowl";
     "beat well"
   end;
```

Each **while** statement contains a continuation condition and a body. The continuation condition is the expression following the **while** keyword, and the body is the statement or compound statement that follows the keyword **do.** When a **while** statement is executed, the body is executed repeatedly as long as the continuation condition remains true. The continuation condition is checked *before* each execution of the body, including the first one, which means that the body won't get executed at all if the continuation condition is false to begin with. This is different from the **repeat** statement—there, because the stopping condition is checked *after* the body is executed, the body is always executed at least once. The difference is subtle, and in the preceding example it makes no difference. But the difference can be important. Suppose, for example, that I have only $50 in my bank account. Consider the difference between

```
repeat "pay John $1 million";
       "subtract $1 million from John's account"
  until "John has less than $1 million in his account";
```

and

```
while "John has more than $1 million in his account" do
  begin
     "pay John $1 million";
     "subtract $1 million from John's account"
  end;
```

The difference is exactly $999,950 in my favor, and it shows how drastic the effect of a subtle difference in programming language semantics can be. Subtle differences like this one are a common cause of bugs.

Labyrinths Are Easy to Make and Hard to Follow

Conditional statements and repetitive statements control the sequence of execution, and for this reason they are referred to collectively as **control structures.** Every programming language has its control structures, some more elaborate than others. For historical reasons, most programming languages have in common at least two basic control structures. One of them is easy to overlook— it's just the "default rule," stating that a sequence of statements is executed in order of appearance unless an explicit control transfer takes place. The other basic control structure states explicitly what to do next—it provides for the unconditional transfer of control to any desired statement, and it's known almost universally as the **goto** statement, spoken and read as "go to." In Pascal's **goto** statement, for example, the label of the next statement to be executed is given immediately after the keyword **goto,** as in the following fragment of Pascal pseudo-code. (In Pascal, a statement's optional label precedes the statement and is separated from it by a colon.)

```
    goto 20;
10: "eat a piece of cake yourself";
20: "give rest of cake away";
```

Unless someone gives part of the cake back, the result of executing the first statement is one less piece of cake for you.

I didn't mention the **goto** until now for two reasons. First, although intuition may suggest that it's obviously needed, in fact it is not, and I wanted to develop the algorithm for Aunt Martl's Sachertorte without it. Second, there's considerable evidence that indiscriminate use of the **goto** is both hard to resist and harmful. As a control structure in modern programming languages, the **goto** has fallen from grace, and for good reasons.

To clarify some of the charges against the **goto,** here is a new version of Algorithm AMT-3 (page 136), written with **goto** and **if-then** statements, and without any **if-then-else, repeat,** or **while** statements—Algorithm AMT-4:

```
begin
      "separate eggs";
      if "torte includes butter" then goto 10;
      "beat egg yolks";
      "beat in the sugar gradually"
      goto 30;
10:   "bring butter to about 70° F.";
```

```
       if "room temperature is very hot"
           then "put the bowl in a deep pan of 60° F. water";
       "beat butter in bowl until smooth";
       "beat in the sugar gradually";
20:    "add egg yolk to bowl";
       "beat well";
       if "no more egg yolks" then goto 20;
30:    "beat in flavorings and dry ingredients gradually"
       "beat egg whites";
       "fold in egg whites";
       "bake cake"
  end;
```

In comparison with Algorithm AMT-3, most people find this version harder to read. If you happen to look at the statement "beat in the sugar gradually" (just before statement 20) and ask what circumstances can cause its execution, the answer requires that you remember or trace through the sequence of events that would cause control to arrive at this point. If you look at the same statement in Algorithm AMT-3, you see immediately that it will be executed only if "torte includes butter" is true.

Careless use of the **goto** leads easily to convoluted patterns of control flow that are hard to follow, and our ability to remember long sequences of events is limited (remember the magical number seven). With **goto**-less algorithms like AMT-3, all the information about control flow is apparent from the visible, static structure—we don't have to memorize sequences of events. Programs that use the **goto** sparingly and with care are easier to read, easier to understand, and less likely to contain subtle errors. In fact, when a friend of mine looked at Algorithm AMT-4 in an earlier draft, he found three bugs! I fixed two of them, but left the third one in as an additional illustration of how easily the **goto** can mislead.

When I explained the concept of subroutines, I mentioned that their use can make programs easier to read and understand. Hence, Algorithm AMT-2 is a little easier to read than AMT-3. That example of differences in readability was the tip of the proverbial iceberg, and in comparing AMT-3 with AMT-4 we see a little more of it. It is a large iceberg, fully capable of destroying nuclear reactors and other modern-day *Titanics*.

The Hardware Life of Software

For years, I had a limited view of computer programming. I usually understood the text of my programs—that is, I understood how to use a programming language as a notation for expressing algorithms, much as I've been using Pascal as a notation in this chapter. But beyond that, how software "worked" was a mystery. I understood fairly well the syntax and semantics of my programs, but I had no idea what computers actually did with these programs.

As long as my programs were correct, and as long as I could depend on the computer to produce accurate results, I didn't need to know what happened inside. But these premises were often false, and I soon realized that programming was easier for those who understood what the computer did with their programs. Such knowledge was particularly important when there were errors to be tracked down, which was most of the time. So I learned more about the various transformations that lead from a written program to printed results. Since then I've found that this information makes it easier to use computers in general, whether I'm programming or using a word processor.

As it changes the values in a spreadsheet, checks for spelling errors in a word-processing document, flies a plane, or controls a nuclear reactor, a computer's behavior often seems complicated. But it's important and helpful to realize that the behavior in fact results from a simple set of instructions, executed obediently by a CPU. The instructions are also executed quickly; even a slow computer can execute a hundred thousand instructions every second. The results of a hundred thousand simple operations cannot always be described simply, which is one of the main reasons why a computer's behavior can seem so complicated and capricious. Another reason is that there are many layers of software between us and the CPU. These layers, which developed more or less historically from the CPU outward, ease the job of writing large, reliable, and powerful programs.

I'll discuss the various layers in terms of a simple example: a program that subtracts two from a number that's stored in one memory location, stores the result in another location, and then chooses the next instruction to execute depending on whether the result of the subtraction is less than zero. I'll follow the historical path, starting with the instructions executed directly by the CPU, and ending with the Pascal code that could lead to those instructions.

Machine Language

The bottom layer of software comprises **machine language,** a term that refers to the patterns of bits that the CPU recognizes as instructions. These instructions define the **primitive** operations of the machine—they are the basis for constructing all the software that runs on the computer system containing the CPU. From the CPU's point of view there is only machine language, and the CPU is always executing a machine-language program. It does so by repeatedly doing two things: executing the current instruction, and determining which instruction to execute next.

The computer is an information-processing machine, and as such it moves and transforms data according to specified procedures. These characteristics are reflected in the CPU's **instruction set,** a term that refers to the entire collection of machine instructions a particular CPU is designed to execute. Some of the instructions, called **transfer** or **move** instructions, do nothing but move data around—among main memory locations or between main memory and registers, those special memory locations contained within the CPU. Other instructions, called **arithmetic** and **logical** instructions, combine and transform bit patterns in various ways. Still other instructions, called **control** instructions, do nothing but determine which instruction gets executed next. Depending on the particular CPU design, some instructions may combine more than one of these general functions.

The bigger, the faster, and the more powerful the computer, the more elaborate its instruction set tends to be. The instruction set of the CPU in a desktop personal computer like the IBM PC has about seventy-five instructions. The CPU of a large computer system—the kind that fills a room—may have several hundred instructions.

Users like you and me rarely see machine language. But every command we issue is interpreted and carried out by machine-language programs. Here's the subtraction example as a machine-language program—four instructions that could be found in eleven contiguous bytes of main memory (reading left to right, each byte shown separately). Each of the four instructions is on a separate line:

```
10100001   00000000   11111111
00101101   00000000   00000010
10100011   00000000   00111111
01111100   00010011
```

The meaning of such bit patterns is specific to a particular CPU design. These happen to be instructions for an Intel 8088—the CPU of many popular personal computers, including the IBM PC. For a different CPU, the same bit patterns would either be meaningless instructions or instructions that don't correspond to the subtraction example. As Intel 8088 instructions, however, these bit patterns have the following meaning (take a deep breath): The first instruction orders the CPU to copy the two bytes (16 bits) starting with main memory location 1111111100000000 into a special CPU register called the **accumulator**—so named because it is used to accumulate the results of arithmetic operations. The second instruction orders the CPU to subtract 2 from the number stored in the accumulator. The third instruction orders the CPU to store the result of this subtraction in the two bytes starting with main memory location 0011111100000000. Finally, the fourth instruction orders the CPU to examine the sign of the result that was left in the accumulator (the result of the subtraction), and to skip forward 19 instructions if the sign is negative. Of the four instructions, only the fourth instruction is a control instruction. This reflects the common convention that the next instruction to be executed is the one immediately following the current instruction, unless a control instruction specifically dictates otherwise.

Assembly Language

What a drag: machine language is obviously cumbersome. Bugs are easy to create but hard to find. Historically, these disadvantages resulted in the development of **assembly language,** so named because it permits one to build (assemble) a machine-language program using a more symbolic notation. In 8088 assembly language, the subtraction example looks like this:

```
MOV   AX,T
SUB   AX,2
MOV   S,AX
JL    N50
```

Each line above is equivalent to the corresponding line in the machine-language version. Notice first that there are symbolic, mnemonic names for each instruction: MOV for "move," SUB for "subtract," and JL for "jump if less than." Symbols are also used for the accumulator, for the main memory locations, and for the destination of the jump instruction. The symbolic name AX always (in the 8088) refers to the accumulator. Symbolic names for the main memory locations, T and S in this case, are chosen by the programmer. In these terms, the instruction sequence means the following: Move the contents of memory location T into the accumulator, subtract 2 from the accumulator, move the result to memory location S, and jump to the instruction labeled N50 if the result of the subtraction was less than zero.

In comparison to machine language, assembly language is easier to read, easier to write, and easier to change. These advantages derive mostly from the use of symbolic names. Symbolic names for data locations are particularly important; they permit programmers to say what should happen to data without saying where the data will be located. These concerns can change independently during software development, so it's important to have software development tools that help to separate them. Assembly language is one such tool.

As in the subtraction example, there's usually a one-to-one correspondence between assembly-language instructions and machine-language instructions, so the translation of an assembly-language program into a corresponding machine-language program is a straightforward process. In almost all cases it's done automatically by a special computer program called an **assembler.**

Programming Languages

So-called **high-level** or **high-order programming languages** take the progression from machine language to assembly language one step further by permitting you to express computations in even more concise and familiar terms. Some popular high-level programming languages are FORTRAN, BASIC, COBOL, Pascal, Logo, and Lisp. Although programs are also written in machine language and in assembly language, the term **programming language** by convention is reserved for high-level languages.

In a Pascal program, the subtraction example might look like this:

```
S := T - 2 ;
if S < 0 then goto 50;
```

The symbols S and T are referred to as "variables". Generally speaking, **variables** in programming languages can be thought of as named "information holders" whose contents can vary. In this particular example, the symbols S and T would have been declared to be the names of integer variables. The ":=" sign in the first statement indicates **assignment**—the statement assigns to the variable S the result of subtracting 2 from the value of the variable T. You can read the second statement by saying "if S is less than zero, then go to statement number 50". Remember that 50 is a symbolic label for another Pascal statement elsewhere in the program.

The values of S and T are single numbers. Other variables can have non-numeric values, such as text characters. High-level programming languages also make it possible to refer symbolically to lists of numbers, strings of characters, as well as to other data collections chosen for convenience in storing and manipulating information. Such collections of data are called **data structures.**

The advantages of a high-level programming language over assembly language are similar to the advantages of assembly language over machine language. Compared to programs written in assembly language, programs written in a high-level programming language are easier to read, easier to write, and easier to change. Fairly complicated calculations can be expressed with a single programming-language statement that uses relatively familiar notation, as in a Pascal statement that computes the total price of a purchase given the number of units purchased, the unit price, and the discount expressed as a percentage:

TOTAL := QUANTITY*PRICE*(1 − DISCOUNT/100)

(Pascal uses "*" as the symbol for multiplication and "/" as the symbol for division.)

A program written in FORTRAN, BASIC, Pascal, Logo, or any other programming language isn't executed directly by the CPU in a typical computer. The program is just data for other programs in the layers of software that exist between the high-level program and the CPU. These other programs deal with the high-level program in one of two basic ways. In one way—the typical way for FORTRAN and Pascal programs—the program is first translated into an equivalent assembly-language program by special programs called **compilers.** The compiler treats the high-level program as data and produces the assembly-language program as its result. In general, a compiler will generate many assembly-language instructions from a single programming-language statement. The preceding Pascal statement, for example, would typically result in 5 to 20 assembly-language instructions, depending on the instruction set of the CPU. The assembly-language program produced by a compiler itself is just data for another program—the assembler—which in turn produces a machine-language program. Although a machine-language program runs directly on the CPU, still other programs are involved in loading the program and getting it started.

The other way in which high-level programs can get executed—the typical way for BASIC, Lisp, and Logo programs—is somewhat more direct. Instead of executing a machine-language translation of the high-level program, the CPU executes a special program called an **interpreter.** The interpreter sequences through the high-level program, one statement at a time. It analyzes each statement, interprets its meaning, takes the required action, and starts again on the next appropriate statement. One way to think of this is that the interpreter program simulates a special-purpose computer that can execute the high-level program directly. Like compilers, interpreters are designed to handle only a specific programming language.

What's the Point?

Assembly language allows programmers to ignore the actual bit patterns that are required to instruct the CPU properly; instead, they can write in terms of symbolic instructions, like MOV and SUB. Programming languages take this a step further by allowing programmers to ignore the CPU's instruction set entirely. For example, if the Pascal subtract-and-test example from the previous section is compiled for the Digital Equipment Corporation (DEC) VAX series of computers, the result would probably be the two VAX assembly-language instructions

```
SUBW3   #2,T,S
BLSS    N50
```

instead of the four Intel 8088 instructions. Although the VAX CPU has plenty of registers, unlike the Intel 8088 its instruction set includes arithmetic instructions that can operate directly on main memory and not just on the contents of the registers. Thus a smaller, faster machine-language program can be generated from the same Pascal source. The Pascal programmer, however, doesn't need to change the original program, or even to know about these differences between the Intel 8088 and the VAX CPU, because the Pascal compiler hides the existence, number, and usage of CPU registers.

Out of the primitive repertoire in the CPU's instruction set, the Pascal compiler builds a repertoire better suited to the expression of algorithms.

The point of all this is to permit programmers to concentrate on expressing the solution to programming problems, while ignoring the details of how the resulting programs actually get executed by CPUs. These are separate concerns, and their separation is achieved by layers of software—including assemblers and compilers—that attempt to insulate programming languages from the instruction-by-instruction operation of the CPU.

The Power of Abstraction

The idea of insulating a high-level programming language like Pascal from differences among the instruction sets of various CPUs illustrates the concept of "abstraction." Abstraction is one of the most important and powerful concepts in modern computer science and engineering, and one that's a lot simpler than its intimidating name might suggest.

What Is Abstraction?

Have you ever borrowed a friend's vacuum cleaner? Did you ask for detailed instructions? More likely you just took it home, plugged it in, turned it on, and vacuumed. When you last rented a car or borrowed one from a friend, did you read the owner's manual? More likely, you just got in, looked around, started it up, and drove away. When you borrow a vacuum cleaner, the characteristics that distinguish a Hoover from an Electrolux are rarely important. And when you rent a car, you're more likely to ask for a four-door mid-size than you are to ask for a particular make and model.

We ignore special cases constantly, and in so doing we exploit general properties. When you pack groceries at the supermarket, you put the six-pack of soda at the bottom of the bag. You do this not because you have a rule for packing soda, but because you have a general rule: heavy things on the bottom. When oil in a frying pan catches fire, you don't stop to consider whether it's peanut oil or safflower oil (let alone the brand), and you don't stop to consider whether the pan is steel, aluminum, or iron. You just put out the fire, using general methods that don't depend on these special-case characteristics.

Whether you realize it or not, when you borrow a vacuum cleaner, rent a car, pack groceries, or put out a pan fire, you are exploiting the power of abstraction. When people hear the word "abstract," they often think of "theoretical" or "abstruse" before their eyes glaze over. Indeed, these meanings appear in most dictionaries, but they usually appear as secondary meanings. The principle meaning of **abstract** is "conceived apart from special cases," and that's the meaning used in computer science and engineering.

A four-door, mid-size car is an abstraction conceived apart from such special cases as Chevy Citations and Buick Skylarks. And a heavy object is an abstraction conceived apart from such special cases as six-packs, quarts, and dog food. We deal constantly with abstractions, although we're more likely to notice them when an abstraction fails, as happened the first time I rented a car that had an ignition key that couldn't be removed without simultaneously pressing a special button. I couldn't get the key out, and I assumed that the ignition lock was broken. This assumption was about to come true when an amused passerby suggested that I press the little button instead of beating on the steering wheel.

Why Are Abstractions Useful?

Given any road map, consider the following procedure: Take a blank piece of paper, draw a little dot for every major town, and label the dot with the town's name. Then, for each pair of towns that are connected directly by a road, connect the corresponding dots and label the connection with the mileage between the towns. The locations of the dots on the paper need not reflect the locations of the corresponding towns, and the lengths of the connections need not be proportional to the actual distances. The resulting drawing is an abstract map that ignores many details of the actual map— details like a consistent mileage scale, the width of roads, and whether roads are twisted or straight.

This particular abstraction happens to be a mathematical object known as a graph; the dots are called nodes, and the connections are called links. Mathematicians have discovered many properties of graphs. For example, there are known algorithms for finding the shortest path between two nodes, and for finding the shortest path that visits a specified set of nodes. Having established a correspondence between a road map and a graph, one can exploit these algorithms to find the shortest route between two towns or the shortest route that includes visits to a specified set of towns (the so-called "traveling salesman problem"). Moreover, these same algorithms can be exploited in solving drastically different problems—problems whose only connection with road maps is the common abstraction provided by graphs. Two examples are finding a good schedule for complicated projects, and finding a good layout for the connections on an electronic circuit board.

Abstractions are useful for two main reasons. First, because an abstraction is in many respects simpler than any of its special cases, it's often easier to reason about an abstraction than about any of the special cases. Second, once results are obtained from studying an abstraction, they can be applied to any of its special cases. Abstractions provide intellectual leverage.

Computers and Abstraction

Instructions in an assembly language are abstractions; they are abstract CPU instructions because they allow the programmer to ignore the details of the machine-language bit patterns. Programming languages like Pascal abstract even further; they allow the programmer to ignore the details of particular CPU instruction sets. Using the definition of "abstract," it's useful to think of languages like Pascal as "conceived apart from special cases," the special cases being the IBM PC, the VAX, and other computers on which Pascal programs can be run; in this sense, Pascal is said to provide an **abstract machine.** Indeed, the design of a programming language can be thought of as the design for an abstract machine, and compilers or interpreters for the language can be thought of as software that implements these abstract machines on particular real machines.

Besides allowing programmers to ignore the details of machine-language bit patterns, assembly language also allows programmers to ignore, or at least to consider separately, the specific locations of instructions and data in memory. For example, in the assembly language version of the subtraction example, two specific memory addresses disappeared in favor of the symbols S and T—think of them as abstract memory locations—and the specific addresses became special-case details handled by the assembler.

In the Pascal version, the abstraction was more general. There, the symbols S and T didn't represent memory locations, they represented integers—mathematical abstractions with well-defined properties. An integer might be represented by the bits in one memory location, as in the assembly-language example, or by the combined bits in several adjacent locations. More than one integer might even be packed into a single memory location. But the integer variables in Pascal and many other programming languages are properly thought of as abstractions conceived apart from these special cases. The abstraction allows programmers to think and write in terms of integers rather than memory locations. Because the abstraction makes it easy to handle a particular class of data, in this case integer data, it's known as a **data abstraction.** Other common data abstractions in programming languages provide abstractions for real numbers, individual characters, and strings of characters.

The process of abstraction doesn't stop with the definition of programming languages—it continues in the programs that are written in these languages. Algorithm AMT-2 on page 133 describes how to "**prepare ingredients**" for *any* torte, not just Aunt Martl's Sachertorte. As such, it is an abstraction that can be applied to different special cases. Moreover, the individual statements in Algorithm AMT-2 are themselves abstractions. There are, for example, several ways to "**separate eggs**", and any of them will do. Such abstractions can be made more specific in later refinements of the algorithm, or they can be left as abstractions by making them into subroutine calls. There are two points: first, that the abstract torte-preparation algorithm is itself built out of abstractions; and second, that we can derive and study the abstract torte-preparation algorithm without considering the special case of any particular torte.

Programmers use abstractions to build other abstractions. When you use a word processor, you don't want to deal in terms of memory locations—you want to deal in terms of characters, words, sentences, and paragraphs. In a successful word processor, the programmer has built data abstractions that allow you to do so. In general, computer programming is properly thought of as the process of designing and building abstract machines—machines for programming, machines for word processing, machines for accounting.

Abstraction is an aid to intellectual control, and it helps us to cope with computers. Indeed, during the last twenty years, abstraction has become one of the key unifying concepts in computer science. Unfortunately, our abstractions often have rough edges that make them hard to handle and flaws through which various special cases poke. Discovering how to describe, create, and manipulate effective abstractions continues to be a challenging and important problem.

Nevertheless, we already do quite well. Abstractions created by layers of software provide the means with which programmers build elaborate and powerful software systems. The results are impressive. When you converse with ELIZA, analyze a spreadsheet, or use a word processor, the computer's capabilities hardly seem consistent with a typical machine-language instruction set. Yet however wondrous and complicated these capabilities may seem, they result from the monotonous execution of billions of primitive operations.

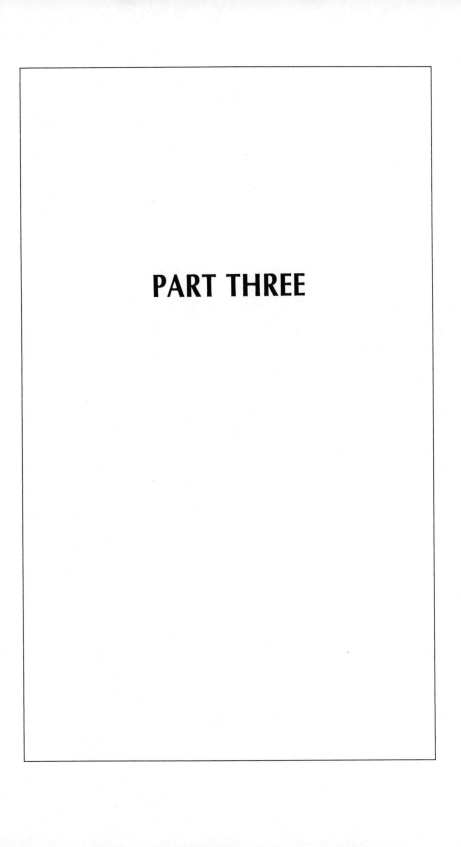

PART THREE

In Parts I and II, I described reactions to computers, interactions with computers, and the internal operations of computers. In Part III, I step back to describe a broader view—one that takes in the reliability, the technology, and the capabilities of the huge computer systems on which our society depends.

Computers are amazing and useful, but they are neither as reliable nor as capable as many people think they are. The main problem is software. The software medium is flexible and powerful, which is a principal reason why computers are so amazing and useful. But the properties of software that lead to these advantages are also catalysts for the production of complexity, and software complexity leads quickly to bugs. Uncontrolled complexity can be a problem in any program, but the problem is most apparent in large programs—programs like those on which banks, nuclear power stations, space shuttles, and ICBMs depend. Uncontrolled software complexity leads to vital computer systems that can fail when we need them most. Uncontrolled software complexity leads to computer vaults that are unsafe for data. Uncontrolled software complexity leads to systems that are easy targets for punk hackers and computer criminals.

In Chapter 8 I discuss the complexity-inducing properties that are at the same time software's biggest advantages and disadvantages, and I discuss how they led to the "software crisis"—the widespread existence of unreliable software. Then, in Chapters 9 and 10 I discuss the modern view of programming that has evolved in response to the software crisis. It is a fascinating and still evolving view, blending elements of literature, mathematics, and architecture to guide the disciplined construction of reliable software. It can be hard to write programs with this modern viewpoint—indeed, the modern viewpoint is based on the recognition that correct software is hard to write—but it's not hard to understand the basic principles and why they are so important.

Finally, in Chapter 11 I turn to another fascinating aspect of large computer systems: their potential for artificial intelligence (AI). I compare the performance of existing computers with that of the human brain, and I explain some of the basic ideas behind AI technology. I also explain why this intriguing technology is less capable and less reliable than the dramatic and well-publicized claims of AI proponents.

CHAPTER 8

Myths of Correctness

The headline summarized a filler item—two column-inches buried within a recent edition of *The Washington Post*. It was the usual story: a department store charged someone too much; the customer complained and was given a credit, but the amount was 29 cents short; the customer complained again and this time received a welcome $161 billion credit. I grinned, clipped the story for my "computer errors" file, and read on. Only later did I notice that the story didn't mention a computer.

We're all quick to blame computers. We do it when we have billing problems, credit problems, delivery problems, reservation problems, and information problems. And if we don't blame the computer, someone else does—usually in a defensive voice over a phone line.

Does the computer deserve all this blame? Literally speaking, of course not. If a machine can ever be said to have caused a problem, the machine is at most a victim of circumstances—people designed the machine, people built it, people chose to use it, and people controlled it. When a car crashes and causes serious damage, we may blame the driver, we may blame some other person at the scene, and we may blame the manufacturer, but we don't seriously blame the car.

Such linguistic distinctions aside, I'm convinced that "the com-

161

puter" often gets blamed when in fact it was either not present at all or present only as an innocent bystander. When we don't get around to doing something or don't want to bother trying, it's easy to say "the computer goofed" or "the computer's down." In a recent newspaper column about these modern excuses, Ellen Goodman cited a friend's comment: "The computer is down" is another way to spell "coffee break." As a modern scapegoat, what could be more convenient and convincing? And satisfying—it's a bit like finding out that your parents are human. Equipped with considerable means, having frequent opportunities, and exempt from the requirement of motive, the computer is an ideal villain. Moreover, when we resolve some mystery by saying, plausibly, "The computer did it," the computer can't defend itself. It is, to use Ellen Goodman's words, the silent butler.

Wrongly accused as it may often be, the computer is responsible for plenty of true horror stories. Indeed, pointing out its occasional and perhaps considerable innocence is a little like absolving Lizzie Borden from an ax murder she happened not to commit. In discussing computer foulups, however, it's important to make those linguistic distinctions that I mentioned before. When we "blame" a computer for some problem, we're really saying either that a hardware component stopped working properly or that someone goofed. That someone could be a hardware designer, a hardware builder, a programmer, or a user.

Hardware components do fail, but modern hardware is remarkably reliable; in a well-established computer system, hardware failures are a relatively infrequent cause of problems. And hardware designs do contain flaws, particularly when they're new designs. But it's fair to say that hardware design flaws are also a relatively infrequent cause of problems, especially in well-established systems. When a computer crashes, chances are that someone tripped it, either a programmer or a user.

I've already discussed how easy it is for novice users to cause problems, especially when confronted with inadequate user-interfaces. Veterans have similar problems, but they often downplay the importance of user-interfaces and blame their computer problems on software bugs; the bugs, however, are not always at fault. Recently I tracked down a variety of stories about computer problems that supposedly were caused by software bugs, and a surprising number of them turned out to be caused by incorrect program operation rather than incorrect programming. You may recall talk of a computer problem during the final moments of the first moon landing—the computer became overloaded and the software responded by restarting itself several times. It turned out that an astronaut had obeyed an incorrect checklist and left a radar switch in the wrong position, thereby generating 13 percent more load on the computer than had been anticipated. This was not the only such problem. Moreover, the astronauts proved themselves capable of making mistakes without being told to do so. Indeed, one study of the computer problems that were encountered during the Apollo space-flight program concluded that about 75 percent of the problems were caused by operator errors. Translated roughly into computer-speak, this means that the Apollo software wasn't user-friendly.

The right stuff, it appears, is no substitute for a good user-interface. Unfortunately, while a good user-interface can protect us from our own errors, we still have to suffer from the other main cause of computer problems: programming errors. And what a cause it is—most large computer programs, even well-established ones, are plagued with bugs.

The Software Crisis

It sounds so melodramatic—the kind of phrase that might be coined by the news media and snickered at by software experts. To the contrary, the phrase arose in the software community ten to fifteen years ago, and it appears regularly in academic textbooks. Indeed, the increasingly apparent difficulty of writing large, reliable computer programs led not only to the term "software crisis," but to a whole field of study, called **software engineering.** Within the computer industry today, the software crisis is widely recognized and widely battled, except by a few who accept it as a fixture along with cancer, venereal disease, and budget deficits.

When people refer to the software crisis, they have in mind several trends. While hardware costs have plummeted, software costs have not. Ten years ago, hardware made up 80 percent of the total cost of a typical large computer system. Today, software makes up 80 percent of the total cost, and this dominant cost can not be predicted well. On top of these financial problems, software usually takes longer to finish than was promised; and when it is finished, it's bigger, slower, and less capable than was promised. These deficiencies might be bearable if the resulting software were reliable, but it isn't. It tends to fail often, and efforts to fix it are just as costly and error-prone as its original development. Moreover, software is hard to change—efforts to improve software capabilities are even more costly and error-prone. While these problems tend to be most obvious in large software systems built for the Department of Defense, they are well known in commercial systems as well. There are, of course, exceptions, but the general trends are clear and widely recognized.

Of all the problems inherent in the software crisis, the most difficult, troublesome, and dangerous is our inability to write error-free software. Two examples may help to make this point.

An Election Night Story

Jimmy Carter conceded the 1980 election before the polls closed in California. His pollster Pat Cadell had warned him that he was going to lose, and the TV networks—using computer predictions based on early returns—declared Reagan the winner early in the evening. Shortly thereafter Carter made it official. West Coast citizens felt cheated. And Democratic candidates on the West Coast, especially those who lost the election, were furious. They believed that the early concession had kept Democratic voters at home and Democrats out of office. Someone even introduced a bill in the U.S. Congress that would make it illegal to concede before all the polls were closed.

It's a tough issue. It's hard to make a strong case that Carter's concession changed the outcome of any particular West Coast election—the races just weren't that close; besides, didn't Republicans stay home too? But the events were still disturbing. The technology of time zones, telecommunications, and computers might have distorted the will of the people. Moreover, this could have happened even if Carter hadn't conceded. After all, people still would have watched TV. They still would have seen those checkmarks next to Reagan's name on the network tote-boards, and they still would have heard well-known correspondents reporting on and interpreting a computer's landslide projections.

These events provide yet another example of our tendency to trust technology. People question how technology is used in elections, they question its effects, but they rarely question its validity. The election process is affected by instant information—information that's gathered, analyzed, and distributed faster than the sun sets. People question the effects of this instant information, but they don't question the information itself.

They should. Consider what happened during the 1981 provincial election in Quebec, Canada. The two main parties in Quebec are the Parti Quebeçois (PQ) and the Liberal Party. There is also the Union Nationale, a small splinter party, and the Marxist-Leninist party, which has only a few hundred members. On election night two TV stations provided coverage. Viewers of one station saw unsurprising results—the Liberals and the PQ led almost all the races (Canadians call them ridings), and the station declared at 8:45 P.M. that PQ had won a majority government.

Things were different on the other station, as well as on a co-operating radio network. The Union Nationale, having been given essentially no chance by anyone, was leading 19 ridings, and a Marxist-Leninist candidate was leading one. The PQ was leading 20 ridings, and the Liberals only 9. The results were astounding, and the commentators lost no time in explaining them. One person dismissed sarcastically "the so-called experts and commentators who had written off the Union Nationale"; the experts were wrong— "the people have spoken."

This apparent upset by the Union Nationale was not only astounding, it was wrong. Software bugs were attributing votes to the wrong candidates, and the TV station went right along. Eventually they admitted the mistake—twenty minutes after the other station had declared a PQ majority—and they subsequently filed a million dollar lawsuit against the company that had assisted them in producing the faulty software. The postmortems were merciless—a columnist in *La Presse* cried "shame . . . dishonor . . . humiliation." A *Montreal Gazette* columnist likened the election night show to *Monty Python's Flying Circus*, citing "general agreement that the election night show was 'the greatest fiasco' in Quebec TV history."

Irony On Board the Space Shuttle

One of the best examples of the software crisis is carried on board every U.S. Space Shuttle Orbiter. NASA has stringent and effective reliability policies. As a result, all of the critical systems in the Orbiter are redundant in one way or another. In the case of data processing, the redundancy is achieved by five identical computers. Four of them are arranged as a voting group—during critical flight periods, all four run exactly the same program and compare their results. Any computer that disagrees with the others is immediately switched out, a procedure that protects the Orbiter against computer hardware malfunctions. The fifth computer operates more or less independently, executing a different program, written by a different contractor. This program provides a backup flight-control system for use if all four voting computers fail.

All four voting computers could fail simultaneously if all four coincidently suffered hardware failures or if there was a fatal bug in the program that they all run. But the possibility of simultaneous hardware failures is not the reason for having the fifth computer run a different program—if it were, more overall reliability could be achieved by having all five computers run the same program while voting. Rather, the existence of independent backup software is motivated by the possibility of a fatal bug in the primary flight software. It's an expensive precaution—by the time the space shuttle program is over, millions of dollars will have been spent on that backup program. Those millions will be spent because NASA recognized that it couldn't develop error-free flight software. That backup computer is a running symbol of the software crisis.

As it happens, the backup computer was involved in a famous bug. While the backup computer operates more or less independently, it is in fact connected to the other four. It "listens in" on them so that it has up-to-date information in case it has to take over. When the backup flight software begins running on the backup computer, it has to synchronize with the primary flight software before it can begin to listen in. This is a bit like "tuning in" to a dance rhythm before trying to follow a dancing partner, which can be tricky. It's particularly tricky for the Orbiter software because the operating systems for the primary and backup flight software exercise control according to different philosophies. The primary software is asynchronous or **priority-driven,** which means that it pays attention to tasks on demand and in accordance with announced importance—it acts like teachers in a progressive day care center who interrupt what they're doing in order to pay attention to the loudest screaming child. The backup flight software is synchronous or **time-slotted,** which means that it pays equal-length attention to tasks in a fixed sequence. It acts like the stern moderator of a debate among eight political candidates, disallowing interruptions and giving equal time to each candidate in turn regardless of how much they have to say or how loud they say it.

Under these circumstances, arranging for the backup software to "tune in" resulted in a bug. As far as bugs go, it was relatively minor, in the sense that it was relatively easy to fix. But you can't fix a bug that you don't know about. This particular bug was extraordinarily subtle, and it remained unknown for a long time. When it finally transpired, it became perhaps the most widely known bug in history, but until then it lurked unknown, waiting for an unlikely but possible series of coincidences. Finally they occurred, not during testing, but at T minus 20 minutes during the countdown of the space shuttle's first scheduled launch, when the backup flight software was turned on. It turned on, but it didn't tune in. The launch was dropped for the day.

The bug didn't prevent the primary flight software from working properly. And the bug wouldn't have prevented the backup flight software from working properly, had the backup software been able to tune in. The bug arose from combining the primary software with backup software that exists only in case the primary software doesn't work. How ironic. And how illustrative of the difficulties inherent in software engineering.

The election night story and the space shuttle story are just two examples of the software crisis, but they're not exceptions. Both of these programs are almost certain to have additional bugs, and most other large computer programs are also erroneous. This fact surprises many people, especially since it's common to hear that a computer program has passed from "development" into "maintenance."

The Meaning of Maintenance

Hardly a week goes by without our being treated to a prediction of how the onslaught of computers will affect employment. It's easy to get the impression that soon we'll all be programming computers or serving fast food. Such predictions are naive, unrealistic, and myopic, but secretly I relish them because they put me in the category of those who can choose what to do. You might think that programming would win hands-down, but there are some distinct advantages to serving fast food. When you slap together a hamburger, for example, at least you know when you're done. Not so with software development—permit me the following distortion of a well-known saying:

> A fast food job lasts from bun to bun,
> But a programming job is never done.

It's obvious that a programming job can be a long one, especially if the object is to develop a large software system, but many people don't realize just how long it can take. One reason for this is the term "software maintenance"—a common linguistic obfuscation that is itself an amusing symptom of the software crisis.

When I mentioned at the beginning of this chapter that we don't seriously blame a car when it crashes, I left out one legitimate target for blame: insufficient or faulty maintenance. Some car parts need occasional attention to keep them working properly, and others need occasional replacement. If these needs are neglected or improperly met, a crash may ensue.

Can insufficient or faulty maintenance be blamed for a software crash? No, at least not in the same sense. Cars and other physical products need maintenance because they wear out with use. But software has no physical properties, only logical properties. When you run the same program a thousand times, some of the computer hardware components may wear and require maintenance, but the software can't wear out. If I made Aunt Martl's Sachertorte a thousand times, I might wear out my kitchen, myself, and my friends, and I might require a hundred fresh copies of the recipe, but the recipe itself couldn't wear out.

A program can work properly a thousand times and fail suddenly the next time. It might thereby give the appearance of having worn out, but what really happened is that an unusual set of circumstances was encountered for the first time. A program doesn't fail because you wear it out; it fails because it didn't work properly to begin with and you finally have occasion to notice that fact. Software is one of the few products of engineering that can truly be said to last forever. This is a marketing dream; unfortunately, few software products are good enough to take advantage of it.

Although the term "software maintenance" is inappropriate, it's common to hear it. Indeed, when most customers buy a large program, they usually enter into a maintenance contract. In return for a monthly or yearly payment, the customer receives occasional revisions of the program. The revisions contain fixes for some of the bugs that were in the original product and perhaps some improvements. The maintenance contract may also provide access to consultants who can help customers work around bugs that they encounter. To a large extent "maintenance" is a software euphemism for "continued development." This language doesn't fool anyone, but it's a convenient fiction. It allows software developers to declare, in analogy with equipment contracts, that a software product has been "developed" and is ready to enter service and be "maintained." If software products were never accepted until they were error-free, few companies could ever finish their development contracts.

If you buy a new car and spend the next year having the dealer fix things that didn't work right to begin with, you don't say that your car is being maintained; you say you bought a lemon. By this criterion, most software products are lemons.

Why Is Software So Hard?

If a car manufacturer produces mostly lemons, we judge the manufacturer to be incompetent, a judgment supported by the existence of many well-built cars. But there is not much well-built software anywhere, and from the bitter taste of a software product, it's unfair to conclude that the builder is incompetent. In general, we know more about building cars well than we know about building software well. What makes software so hard? The short answer is "complexity," but the question deserves a slightly longer answer.

The Inadequacy of Testing

A common reaction to the story about the space shuttle bug is surprise that the bug wasn't caught during testing. Doesn't NASA conduct exhaustive testing before they try to shoot off a rocket with some men in it? Are the software-testing personnel incompetent, the testing procedures inadequate?

The answers to the last questions are "no" and "yes": No, the software-testing personnel are *not* incompetent; and yes, the software-test procedures *are* inadequate. The inadequacy of software testing is not the result of incompetence. No matter how competent you are, it's impossible to expose all of the bugs in a program by means of exhaustive testing.

This fact does not apply just to large-scale software. You hold a more down-to-earth example in the palm of your hand every time you use an electronic calculator. These calculators are popular and convenient, and they usually display correct results. But not always. Indeed, most popular calculators display wrong answers some of the time. An American mathematician and computer scientist, William Kahan, has documented, studied, and helped to correct such problems for years. Here's one of his examples, a compound interest problem for financial calculators. The problem is called "A Penny for Your Thoughts."

A bank retains a legal consultant whose thoughts are so valuable that she is paid for them at the rate of a penny per second, day and night. Lest the sound of pennies dropping distract her, they are deposited into her account to accrete with interest at the rate of 10% per annum compounded every second. How much will have accumulated after a year (365 days)?

You can solve such problems with financial calculators just by pressing a few keys. In this case Kahan did so using ten different popular calculator models. He got four different answers:

$331,667.00
$293,539.00
$334,858.18
$331,559.30

In calculator software as in space shuttle software, errors can result from bugs that lead directly to wrong answers or from the gradual buildup of inaccuracies in repetitive calculations. But, whatever the cause of the errors, it isn't possible to find all of them by testing.

To see why, let's suppose that you've accepted a job with Calcutronics—a new company that makes electronic calculators. Your first assignment is to evaluate the prototype of a new model, a standard calculator that handles eight-digit numbers, and see if it works correctly. Like other electronic calculators, the Calcutronics model is really a small computer. When you press a key to multiply, divide, or compute compound interest, your action invokes a program stored in read only memory (ROM). Even the addition key initiates a program; the CPU does have an **ADD** instruction, but it only adds one-digit numbers. The addition of two eight-digit numbers is accomplished by a program.

How should you go about evaluating the calculator? One approach is simply to try every operation and see if the calculator gives the correct results. This approach is called **black box testing**, probably because black boxes are opaque and their contents have to be evaluated by means of externally observable characteristics. Because black box testing focuses on the calculator's actual behavior and ignores its internal design, it seems to be the least biased and most reliable way to proceed. Unfortunately, you would never finish.

Consider just the addition function. To test it exhaustively, you would have to add every possible combination of eight-digit numbers. Depending on how you like to see it written, there are 10^8, or 100,000,000, or one hundred million different eight-digit numbers. As for the number of different combinations of two eight-digit numbers, there are 10^{16}, or 10,000,000,000,000,000, or ten million billion of them. So to test addition, you would have to add ten million billion pairs of numbers. You might be tempted to cut this in half since, for example, $8+2=2+8$. But $8+2$ and $2+8$ are not the same operations from the calculator's point of view. The difference is just the order in which you enter the two numbers, but that might result in their being handled differently, so you have to try both cases.

How long the complete test would take depends on how fast you work, but it's doubtful that you could test more than one addition per second. If you did one per second, forty hours per week, year round, it would take about 1,300 million years to finish, give or take a few million years. You might be able to save time by building a machine that worked around the clock, punching in numbers and photographing the answers for you to check during the day. Such a machine would also take about one second per operation, but its constant attention to the task would reduce the total time a lot. With the machine, you could finish in about 320 million years.

You could bypass the mechanical operation of the keys, connect a computer directly to the electronics inside the calculator, and use the computer to try all the different combinations. But you still wouldn't be done in a reasonable amount of time. Even if the computer could test one million additions per second (faster than most calculators would allow), the total time would be about 320 years. Job security, perhaps, but not exactly what Calcutronics was looking for.

It might seem reasonable just to test the program with selected input values scattered throughout the overall range and conclude, if the program works properly for these values, that it also works properly for every value in between. This approach could be valid if a large number of input values were tried, provided that the addition process is continuous—i.e., provided that a small change in the inputs corresponds to a small change in the output. But we can not assume continuity. For example, a program can easily treat one particular input value in a totally different manner than it treats nearby values—all it takes is a single **if-then** statement. And if you view the program as a black box, you can't rule out the possibility of such **if-then** exceptions being inside. Just because the result of

32000 + 767

is correct doesn't mean that the result of

32001 + 767

is likely to be correct.

The calculator's addition program had only two eight-digit inputs, yet the total number of input possibilities made black box testing impossible. It wasn't even close. Black box testing would be exhausting but not exhaustive—there just isn't enough time. Furthermore, most programs have even more input possibilities. The conclusion is inescapable: black box testing is hopelessly inadequate as the basis for any thorough evaluation of computer program correctness.

This conclusion and the arguments that lead to it are summarized succinctly by a remark made by the Dutch computer scientist Edsger W. Dijkstra. It is probably the most often-quoted statement in the computer science literature:

Program testing can be used to show the presence of bugs, but never to show their absence!

No matter how diligently you test computer programs, you cannot test them completely. The point is really that simple. And that profound.

That program testing is inadequate doesn't mean that we should eschew it—not to test a program before depending on it would be foolish. But the inadequacy of testing does mean that the role of testing is corroborative rather that definitive—testing can corroborate our belief in the correctness of a program, but we must have other evidence. Where can that other evidence come from?

If black-box testing is inadequate, the only alternative is to open the box and examine what goes on inside. It follows that evidence for a program's correctness must somehow involve not just running the program, but studying the program itself. Stated differently, our belief in the correctness of a computer program must arise substantially from intellectual arguments based on the program's written text. So far, so good. Unfortunately, if we cannot understand the program's text, we certainly cannot argue effectively about its correctness, and therein lies a major problem.

The Curse of Flexibility

Fluency in English doesn't guarantee that you can understand every document that's written in English. It all depends on the clarity of the writing. Not so long ago, I struggled with a particularly unpleasant tax form. Neither my educational background nor my ego encouraged me to give up, but after three hours I did. Unfortunately I still had to fill out the form, so I called the special phone number that the IRS maintains for citizens who need help. Eventually I spoke to a real, live IRS agent and explained my problem. Almost immediately, he laughed. "Oh that form! Listen, nobody understands that form. Here's what you do. . . ."

The situation with many computer programs is similar. A few years ago some colleagues and I embarked on a software-engineering project in which we proposed to demonstrate various modern methods in a practical way by rebuilding the obsolescent flight program of a naval aircraft. In the preface to our first report on the project, we mentioned that the existing software, among its other faults, "is not fully understood by the maintenance personnel." This statement got us into trouble— not with the maintenance personnel, who agreed with us emphatically, but with their project sponsors, who happened also to be our sponsors. I never regretted the statement, however. It describes accurately the program in question, and many other programs as well.

There are many reasons for programs being hard to understand, but at the root is the mixed blessing of flexibility. A computer's behavior can be changed radically by changes to its software. Whether a programmer wants to fix a bug, change an existing function, or add a new function, it's easy to bend the software accordingly. In principle, this flexibility is a blessing; major changes can be accomplished quickly and at low cost. But the blessing is hard to receive. Because it's easy to make changes quickly and without considering all the ramifications, software complexity can grow quickly, leading to software that's hard to read, hard to understand, likely to contain more errors, and likely to require further modifications. These results are hard to avoid, and their effects can be crippling. In this light, the computer's flexibility looks less like a blessing and more like a curse.

The computer's flexibility is unique. No other kind of machine can be changed so much without physical modifications. Moreover, drastic modifications are as easy to make as minor ones, which is unfortunate, since drastic modifications are more likely to cause problems. With other kinds of machines, drastic modifications are correspondingly harder to make than minor ones. This fact provides natural constraints to modification that are absent in the case of computer software. Such natural constraints bring discipline to machine design and construction as well as to machine modifications. In the case of airplane construction, for example, feasible designs are governed by the mechanical limitations of design materials and by the laws of aerodynamics. There is a resulting, nature-imposed discipline to the design process that helps to control complexity. In the case of software construction there are no such limitations or natural laws. This makes it too easy to build enormously complex software. Indeed, the structure of typical software systems makes the humorous complexities of Rube Goldberg's fanciful machines look elegant in comparison.

The flexibility of software also encourages the redefinition of tasks quickly, often, and late in the development process. Imagine that General Motors is 90 percent finished with a new car development; deliveries are scheduled to begin in a few months. Suddenly it's decided to have the driver control the car from the back seat and to include a separate air conditioning system for the trunk. It's a laughable scenario, but analogous ones are common in software development. Software is the resting place of afterthoughts.

Software's flexibility is deceptive and seductive. It encourages programmers to plunge in, and they tend to do so; premature construction is a common software problem. Few would be foolish enough to begin building a large airplane before the designers finished detailed plans. In the case of large computer programs, however, few are wise enough to wait. Because the software medium appears to be so forgiving, it encourages us to begin working with it too soon; we begin our attempts before we understand our tasks. It's extremely hard to build a large computer program that works correctly under all required conditions, but it's easy to build one that works 90 percent of the time. It's also hard to build reliable airplanes, but it's not particularly easy to build an airplane that flies 90 percent of the time.

With software, it's easy to start out and hard to finish. Total success is difficult because the flexibility of software facilitates partial success at the expense of unmanaged complexity. And once a program's complexity has become unmanageable, each change becomes as likely to hurt as it is to help. Each new feature may interfere with several old features. Each attempt to fix a bug may create several more. The feeling of "one step forward, two steps back" is a common one. The programmer facing poorly understood, overly complex software is like Brer Rabbit facing the Tar Baby.

Like airplane complexity, software complexity can be controlled by an appropriate design discipline. But to reap this benefit, people have to impose that discipline; nature won't do it. As the name implies, computer software exploits a "soft" medium, with intrinsic flexibility that is both its strength and its weakness. Offering so much freedom and so few constraints, computer software has all the advantages of free verse over sonnets; and all the disadvantages.

Invisible Interfaces

A common and sensible approach to writing large programs is to divide the problem into parts and write a separate program for each part. Unfortunately, to do so is simple in principle but surprisingly difficult in practice. Much of the difficulty arises because the separate programs must interact to solve the overall problem—some programs have to exchange information with other programs, and some programs have to control other programs. These interactions take place across various interfaces that exist among the programs.

Consider once again the magic-trick example. If you enter '0' (zero) in response to the prompt

PLEASE ENTER A NUMBER BETWEEN 1 AND 10:,

the magic trick fails, and the program stops with the explanation

FATAL ERROR . . . REGISTER OVERFLOW AT AF45
712 547 234 232
777 234 342 455
209 487 439 332
>

This error can be viewed as the programmer's fault for not having checked the number that was entered before using it as a divisor. But a subtler explanation is possible if more than one programmer is involved. Suppose that the magic-trick problem were divided into two parts, and that there were two programmers—one responsible for a program that obtains a number from the user, the other responsible for a program that performs the calculations involved in the magic trick. Not knowing that the input value would end up as a divisor, the first programmer may have assumed that the range 1 to 10 was a suggestion rather than a requirement. Meanwhile, the second programmer knew that the input value would end up as a divisor, but assumed that the first programmer wouldn't permit anything outside of the required range 1 to 10. The fatal error arose because the two programmers made conflicting assumptions about the number that was to be passed from one program to the other. This error is a small example of an interface-related bug.

A larger and classic example is provided by the space shuttle bug. Two separate programs—the primary flight software and the backup flight software—were written in order to solve the overall problem of flying the Orbiter reliably. Considered separately, they both worked (at least from the viewpoint of the bug in question), but a bug in their interface prevented the backup flight software from starting up properly.

Here's another example of an interface-related bug, one that I encountered on my Lisa. There are two Lisa programs that I've been using regularly. One is the word-processing program that I'm using to write this book; the other is a telecommunications program that allows me to use the Lisa as a remote terminal connected by telephone to computers elsewhere. Although these are separate programs, they can be used together—you can switch back and forth from one to the other and you can also transfer information from one to the other. For example, you might switch from the telecommunications program to the word-processing program, pick up two or three paragraphs from a word-processing document stored on the Lisa, switch back to the telecommunications program, and transmit those paragraphs to the remote computer. The two programs worked together well until I needed to transmit not just a few paragraphs, but an entire copy of a long document. I successfully transmitted the text in the prescribed manner, but when I switched back to the word-processing program, it crashed.

The crash was inconvenient and somewhat damaging; I lost the most recent changes I'd made to the document I was working on. The crash, however, was graceful, and in this respect it exemplified a good user-interface. For example, the crash was accompanied by a series of error messages that were user-oriented instead of programmer-oriented. The messages spoke of "technical difficulties" (a term that covers a multitude of sins but is generally accurate), they spoke of attempts to recover my document, and they warned me that I might have lost the changes that indeed were lost. And when the crash was finally over, only the word-processing program was affected—I didn't have to restart the Lisa's operating system, and the telecommunications program was still connected properly to the remote computer.

The cause of the crash was a bug in the interface between the word-processing program and the telecommunications program. The two programs made conflicting assumptions about the use of the Lisa's main memory—when I used the word-processing and telecommunication programs separately, the problem didn't arise because the programs didn't have to share the available memory with each other. But when I used them together, they did have to share. It appears that this fact was overlooked somewhere in the telecommunications program. When I transmitted a copy of the long document, the telecommunications program had to use more memory, and it ended up using a portion of the memory that was already being used by the word-processing program. The resulting interference caused the word-processing program to crash.

The more complicated an interface, the more likely it is that something will fall through a crack. Software interfaces are so error-prone because it's so easy to build complicated interfaces. The curse of flexibility strikes again—it's easy to make practically anything depend on practically anything else. Moreover, the dependencies can be subtle, and they're almost impossible to detect by studying the programs involved. For example, one program might work properly only if another program can be relied on to finish its job in a specific amount of time. Indeed, the shuttle bug originated from just such an assumption. About a year before the first attempted launch, a change was made to the primary flight software that caused certain operations to take longer. This violated a subtle assumption that was implicit in the backup flight software, but not stated explicitly anywhere. The resulting bug eventually showed up on launch day. This example shows the need for a broader definition of "interface": namely, a software **interface** between two programs comprises all of the assumptions that the programs make about each other.

Like software, physical machines such as cars and airplanes are built by dividing the design problems into parts and building a separate unit for each part. The spatial separation of the resulting parts has several advantages: It limits their interactions, it makes their interactions relatively easy to trace, and it makes new interactions difficult to introduce. If I want to modify a car so that the loudness of its horn depends on the car's speed, it can be done, at least in principle. And if I want the car's air conditioner to adjust automatically according to the amount of weight present in the back seat, that too can be done—again, in principle. But in practice such changes are hard to make, so they require careful design and detailed planning. The interfaces in hardware systems, from airplanes to computer circuits, tend to be simpler than those in software systems because physical constraints discourage complicated interfaces. The costs are immediate and obvious. No comparable constraints operate in software systems. Indeed, the medium encourages complicated interfaces. The total costs of a software interface, like the costs of a hardware interface, grow quickly with complexity. But the costs of a software interface are neither immediate nor obvious.

Interface related problems are common in any computer program, but their prevalence grows rapidly with the size of the program. In computer programs that are designed and built by a large team of programmers, all sorts of subtle assumptions are made, and there's plenty of opportunity for conflict. This is just one example of the problems that beset programmers when they attempt to scale up their previous successes.

Scaling Up Is Hard to Do

When I was a boy I liked to build model airplanes and fly them at a park near my house. And it wasn't so long ago that a friend and I put together a twelve-inch-long rocket and fired it off gleefully. As small as these achievements were, I'm proud of them. But I've never attempted to build jumbo jets or space shuttles.

Such attempts would be absurd. Yet analogous attempts are common in the case of computer software. Programmers are constantly acting as if the skill and effort required to build small computer programs can be scaled up easily to build large programs. They're wrong.

I'm as guilty as anyone. When I first went to work at the Naval Research Laboratory, I was asked whether I had any experience with simulation programs. Indeed, I had. As part of a political science project that I did while a college senior, I had written a FORTRAN program that simulated the international arms race. I was quite impressed with this project, for which I recall receiving an "A," and I felt confident that I could handle the assignment that was proposed for me: managing the development of a computer program that would simulate in detail the electronic systems aboard large naval ships. That my previous experience consisted of a small program embodying the straightforward computation of a trivial equation gave me only slight pause. After all, the principles were the same—only the size of the program was different.

I accepted the assignment along with someone else's estimate of what it would take: around $250,000 and two years. Both the money and the time seemed excessive to me, but I thought it prudent not to object. Soon thereafter I engaged a willing contractor, and within six months or so I began receiving design documents and programs. Some months after that, it became obvious that the project was in serious trouble. I appealed for more money, which I received and passed on to the contractor, who assigned more people to the task. The trouble got worse.

What I needed then was more knowledge and not more money. Among other things, I wish I had known about Brook's Law:

> Adding manpower to a late software project makes it later.

This is the high-tech equivalent of "too many cooks spoil the broth." It was first stated—at least in this eloquent form—by the American computer scientist Frederick P. Brooks, Jr., a man who should know. Fred Brooks has been described as "the father of the IBM System/360," a series of computers that was IBM's main line from the middle 1960s to the early 1970s. Brooks served as project manager during the development of the system, and he directed the development of OS/360, its software operating system. OS/360 was, by previous standards, a huge undertaking. The project had now-classic problems with time, effort, money, and reliability, and it has become a famous symbol of the software crisis. As Brooks said about just one of his decisions,

> It is a very humbling experience to make a multi-million-dollar mistake, but it is also very memorable.

Like the problems of my simulation project, the problems of OS/360 and countless other projects arose in large part because the people involved underestimated the difficulty of scaling up their previous efforts. Brooks put it this way:

> The second system is the most dangerous system a man ever designs.

He called this the **second-system effect**, and the term has caught on. Brooks was right when he said that "all programmers are optimists," but where does this optimism come from? There is, I think, a basic human urge to accomplish more. In engineering we see this as an urge to scale up previous results. We see it in airplanes, bridges, and buildings. We also see it in software. But the urge to scale up the size or performance of physical objects is tempered by natural and obvious impediments that require technological improvements in making and handling building materials. No matter how much you may want to build a Boeing 747 instead of a Piper Cub, a Saturn Booster instead of a V2 rocket, or the Brooklyn Bridge instead of a stream crossing, the difficulties and the costs of mistakes are sufficiently obvious to serve as effective restraints.

With computer software, it's different. The obvious limitations to performance are the computer's speed and its storage capacity; if both are doubled, the goal of doing twice as much looks easy to achieve with software. But the appearance is deceptive, because the process is highly nonlinear. Even if twice as much is possible when you double a computer's capacity, it becomes much more than twice as hard to succeed.

Scaled-up software is not only harder to produce, it's harder to maintain. Clearly written software is important generally because of the inadequacy of testing. But the importance of clarity grows as software is scaled up because large programs that are used over a period of many years tend to be maintained by programmers other than those who wrote the program to begin with. When you maintain a program you wrote yourself, your understanding of its operation is assisted by your memory. When you maintain a program that someone else wrote, your understanding must rely solely on the program's text.

Increased computer capacity attracts the programmer like a charm, but its effects are tempered by the curse of flexibility. Faster computers and larger memories are natural catalysts for the production of complexity. To manage that complexity, we need appropriate tools—something that's analogous to the materials handling technology used in building physical objects. Such tools are themselves made out of software. Examples include compilers for programming languages, and various other programs that help us to write, analyze, and test software—generically such programs are called **support software.**

Support software helps in managing complexity, but programmers also need help in avoiding excessive complexity in the first place—they need something analogous to the discipline imposed by the natural constraints on building physical objects. Here we must, in the jargon of modern society, resort to "self-help." We must impose our own constraints, our own discipline. We can do so by means of appropriate programming languages, programming methods, and documentation methods. Complexity can't be eliminated, but it can be managed; the software engineer's job is to manage unavoidable complexity and avoid unmanageable complexity.

Advances in electronics technology have made small personal computers inexpensive and widely available. But although their components are modern, their capacities and their software tools often reflect twenty-year-old technology. Formative exposure to these machines has some potential for harm. If people are uninformed and also gain their computer experience by using twenty-year-old languages to write small programs on these small machines, they will learn nothing about those aspects of computers that led to the software crisis, and they may well be predisposed to follow in the intellectual footsteps of their predecessors. Of course, those who go on to bigger machines and bigger programs will soon learn about the problems I've been discussing, as will those who are educated in other ways. The rest, however, stand to be misled profoundly about modern computer technology.

The Myth of Incorrectness

Of all the potential effects of the personal computer craze, the one I find most troublesome concerns attitudes about software correctness. When people start writing their own bug-ridden programs, their experience may temper their harsh reactions to the computer foul-ups around them. People know that to err is human, and they may believe that programming errors are just another example of human fallibility. This is a myth. But it's a myth that will be reinforced if people are exposed to such nonsense as I found recently in the introductory manual of a popular personal computer:

"Bugs," in computerese, refer mostly to imperfections in software. These bugs are minor flaws that the people who wrote the program couldn't foresee.

Statements like this breed familiarity with bugs, but not contempt. Bugs are indeed flaws, but not all bugs are "minor flaws"—some of them can kill people. Moreover, it's not true that people *couldn't* foresee the bugs; what's true is that people *didn't* foresee the bugs.

Programmers do make mistakes, but today's software problems are not statistical evidence for human fallibility, they are the unavoidable results of the programming languages and methods that we use. The complexity of the computer programs we write has grown faster than our ability to write them correctly. This mismatch is one of the most important and difficult technical challenges of our time.

CHAPTER 9

Programming As a Literary Activity

I first learned the rudiments of computer programming in about 1966, when I took that part-time, physics-department job that I mentioned in Chapter 1. At the time many on campus needed to learn how to program, and the computer center's staff responded with an endless series of introductory courses. I enrolled in one that promised to teach the essentials in a couple of weeks of evenings.

It was a course in the programming language FORTRAN, an acronym derived from "**for**mula **tran**slator." In many ways FORTRAN is just that—a convenient means of translating mathematical formulas into computer instructions. It was developed in the mid-1950s by John W. Backus and others at IBM, and it quickly became the first widely used language for scientific programming. It was an enormous step forward. Almost thirty years later FORTRAN is still the most widely used programming language for scientific and many other purposes. Today's FORTRAN incorporates a variety of more recent features, but its essential characteristics are unchanged.

I wish I could say that FORTRAN's continued popularity is an example of something having been done right the first time, but it's more an example of how a new technology can be driven by the momentum of early developments. Ten years ago, Edsger Dijkstra put it well:

If there had been a Nobel prize for computing science, FORTRAN would have been an achievement worthy of it. But that appreciation should not engender the mistaken belief that FORTRAN is the last word in computing; on the contrary, it was one of the first words. It is just no longer adequate: since the twenty years of its existence,

185

the computing scene has changed by several orders of magnitude. How could it still be adequate? We don't control Jumbo Jets by whip and spur!

The computing scene has indeed changed radically since the 1950s—the cost of computers has plunged while speed and storage capacity have soared. Unfortunately, this hardware bliss was tempered by the emergence of the software crisis. Again, Dijkstra:

> As long as there were no machines, programming was no problem at all; when we had a few weak computers, programming became a mild problem, and now that we have gigantic computers, programming has become an equally gigantic problem.

The gigantic problem of software hasn't been solved, but it hasn't been ignored. Indeed, the past thirty years have seen considerable progress, both intellectual and technological. Today we understand better the causes of unreliable software, and we know better how to avoid it. To assist us in trying, we have better programming languages, better support software, and better programming methods.

The achievement of these advances has been slow, but their acceptance into regular practice has been even slower—the continuing popularity of FORTRAN being but one example. The reasons are complicated, but they include the traditional reasons for slow technology transfer: the large amounts spent on existing investments, the cost of new investments, and the human reluctance to abandon the familiar and the habitual. Yet there seems to be more to it in the case of software. In particular, the traditional reasons for slow technology transfer are compounded by two others: First, the voracious optimism of programmers and their managers tends to obscure the need for change. And second, the new software technology requires that programmers change not only what they do, but also how they think about what they do.

One important change in thinking is the recognition that programming is program writing. This vacuous-sounding distinction is in fact responsible for major improvements in software reliability. Enormous benefits arise simply from recognizing that programming is a literary activity.

The Programmer As Writer

I admire good writing. I hope you know what I mean by this—unfortunately, like good music, good food, and other good things in life, good writing is easier to detect than to describe. Roughly speaking, however, by "good writing" I mean clear, concise, elegant, and simple prose.

I haven't always admired good writing. Like others of my generation, I suffered for years under the tyranny of the ten-page paper. Moreover, I attended college during the 60s, when curricula did not revere The Basics. My stylistic conversion came later, after a friend marked up one of my draft papers with questions I hadn't thought of and handed me a copy of George Orwell's essay "Politics and the English Language." I doubted that political writing had much to do with scientific writing, but I was willing to learn how to communicate better. Besides, it looked easy. Only when I started paying attention to my own writing style did I realize not only how hard it is to write well, but also how essential it is for clear thinking and reliable arguments on any subject. Being thus both difficult and essential, good writing may well sound like a recipe for depression. Fortunately, clearer thinking and more reliable arguments can result from attempts to write well, even if the final prose doesn't pass muster.

My experiences with computer programming have been similar. For years I wrote programs without thinking about their style. In fact, their style determined how easy it was for me and others to read and understand them, but I didn't consciously address this fact—I didn't think that style had anything but a mildly artistic relevance. My conversion came when I learned about the software crisis and about attempts to deal with it. The concept of a good style, I realized, is as relevant to computer programs as it is to prose. Just as a good prose style is essential for clear thinking and reliable arguments, a good programming style is essential for clear thinking and reliable software. Like good writing, good programming is hard, but many of the benefits likewise arise from a concerted attempt. Good programming is good writing.

Although it's well known that programmers write computer programs, many people don't think of the programmer as a writer—

they see people as the writer's audience, and computers as the programmer's audience. This picture changes, however, in light of the conclusion that confidence in the correctness of a computer program must arise substantially from intellectual arguments based on its written text. This means that, once you've written a computer program, you must convince yourself that the program is correct by *reading* it. Moreover, if others have to modify your program in order to enhance its capabilities, they must understand the program well enough to make the changes without introducing errors. To achieve such understanding they must *read* your program. People, therefore, are among the audience of computer programs—indeed, it's no exaggeration to say that people are the most important audience of computer programs. To test a program, we run it. But if we are to rely on the program, we have to read it. Programmers are writers.

As writers who produce text for a human audience, programmers must be concerned with those qualities that distinguish good writing from bad. It follows that programmers should try to make their programs easy to understand, which in turn puts a premium on clarity of expression—the text of a program should express clearly the program's meaning. Unfortunately, easier said than done.

The Infamous **goto** *Statement*

Use of the **goto** statement provides a dramatic and famous example of how the meaning of a program can be well hidden by its text. I mentioned the **goto**'s fall from grace in Chapter 7, and I gave a small example of the **goto**'s disadvantages.

The **goto**'s powerful ability to obscure meaning was noted by many observers over a period of years beginning in about 1960, but it became a *cause célèbre* after Edsger Dijkstra published a short technical note in 1968. In his "Go To Statement Considered Harmful," Dijkstra repeated the charges against the **goto**, he pointed out that programmers can do without it, and he argued that we should do more than resist its use—namely, that we should make the **goto** impossible to use by eliminating it from our high-level programming languages.

Dijkstra offered a plausible explanation for the **goto**'s harmful effects by pointing out that our ability to understand a **goto**-ridden program depends on our ability to remember and understand long sequences of events, which is something that most people do not do well. (Have you ever driven into a gas station and asked for directions, only to forget some of them before the attendant even finished talking?) In contrast, our ability to understand a **goto**-free program depends on our ability to analyze the visible, static structure of the program's text, which most people do better. As for the feasibility of doing without the **goto**, this was established by a theorem proved in 1966 by Corrado Böhm and Guiseppe Jacopini. The Böhm-Jacopini theorem stated essentially that all programs can be written with only three control structures: a sequential structure like the semicolon in Pascal, a conditional structure like the **if-then-else** statement, and a repetitive structure like the **while** statement or the **repeat** statement.

The pros and cons of the **goto** quickly became the subject of a widespread controversy. The software crisis was becoming apparent with force, and many saw the elimination of the **goto** as an easy recipe for salvation. Zealots took up the hue and cry, devoting themselves to anti-**goto** pogroms. Nonbelievers dug their heels in— they saw the new religion as unrealistic and too dogmatic.

After some years a moderate consensus emerged. The nonbelievers had to admit that their programs were better off with fewer **goto** statements. And the zealots had to admit that there were cases where an occasional **goto** actually improved a program's clarity. It became clear that, although **goto** could be eliminated from Pascal-like languages, it shouldn't be eliminated—when languages that omitted the **goto** were used for substantial programming, the **goto** soon reappeared in disguise. Even Dijkstra felt compelled to say that he was not "terribly dogmatical" about the **goto**, in itself a noteworthy event.

Let me add that the last word on the **goto** isn't in yet—we still don't understand how best to use or not use it. We know that it's not always beneficial to eliminate **goto** statements by means of Pascal-like **if-then-else** and **while** statements, but the same conclusions may not hold for other control structures. Several new control structures have emerged from research, and they may well lead to successful programming languages that facilitate clarity in all cases without **goto**s. Meanwhile, the **goto** stands not only as an example of the connection between software reliability and programming style, but also as an example of our incomplete knowledge.

Seeing How to Write Clearly

One of the benefits of recognizing the literary nature of programming is that we can turn for inspiration to centuries of experience with natural languages.

The Role of Language

Our thoughts and our language are inextricably bound. Our language affects what we think about, how clearly we think, and how clearly we express ourselves. It is considerations like these that motivated the likes of H. W. Fowler, Sir Ernest Gowers, William Strunk, and E. B. White—all legendary fighters for literary clarity. These fighters and their followers have won many individual battles, but few wars. The general evolution of our natural languages is determined by the tides of less articulate, less deliberate, but stronger and widespread forces.

We have much more control over the artificial languages that we use for computer programming. The evolution of a programming language occurs by the distribution of revised compilers and interpreters, not by word of mouth. Moreover, a programming language arises in the first place as a conscious intellectual product that embodies deliberate choices. But how to choose? Certainly if clarity of expression is one of our goals when writing programs, we must choose our language in part by the extent to which it facilitates clear writing. As George Orwell wrote:

> The slovenliness of our language makes it easier for us to have foolish thoughts.

The choice of control structures is perhaps the best example of how the design of our programming languages affects the clarity of our programs—the harmful effects of the **goto** can be mitigated by providing control structures like **if-then-else** and **while**. I've already explained that these control structures contribute to readability by making the flow of control apparent from the visible, static text of the program, which alleviates the load on the reader's short-term memory. This effect is somewhat analogous to the effect of structure in a literary document; readability is affected by the arrangement of sentences, paragraphs, sections, and chapters. Control structures like **if-then-else** and **while** also contribute to readability in another way, namely by virtue of their simple, easy to understand semantics. For example, the meaning of the **if-then-else** statement is easy to explain, easy to understand, and easy to remember.

The importance of choosing language features with simple semantics extends beyond the choice of control structures. Complicated semantics are hard to remember; they encourage mistakes on the part of writers and misunderstandings on the part of readers. Unfortunately, complicated semantics are hard to avoid. When designing a new language, there's a natural tendency to make it as powerful or capable as possible. Language designers want to make it easy to write a wide variety of programs, and they often try to accomplish this by including a wide variety of language features. By itself, each feature may be relatively simple to understand and may seem like a good idea. But the features often do not fit together smoothly, in which case the combined set of features becomes a jumbled collection rather than a unified whole. Moreover, fitting the features together often requires special treatment for special cases, which further complicates the semantics. The resulting language soon becomes hard to learn, hard to use, and error-prone. The whole is less than the sum of its parts.

One of the best examples of an overly complicated language was developed in the mid-1960s by a committee of IBM employees and IBM computer users. Called PL/I (for Programming Language One), it was intended to be the principal programming language for the IBM System 360 series of computers. Like OS/360, which I mentioned in the previous chapter, PL/I exemplifies the second-system effect. PL/I was intended to replace both FORTRAN, which was popular for scientific applications, and COBOL, which was popular for business applications. By trying to make PL/I well-suited for practically anything, its designers produced an extremely large language with lots of features, options, and special cases. It is, in a way, the jack-of-all-programming-trades. Although PL/I is in widespread use and does have its devotees, it's generally recognized as being much too large and complicated. It's the kind of language that contributes to the software crisis rather than helping to alleviate it. Edsger Dijkstra put it well:

> I absolutely fail to see how we can keep our growing programs firmly within our intellectual grip when by its sheer baroqueness the programming language—our basic tool, mind you!—already escapes our intellectual control.

You might react to such polemics by thinking about how well we do with English and other natural languages. English is a large language that's hard to learn; its syntax and semantics are filled with special cases and subtle distinctions. Why should we have different expectations for artificial languages?

Indeed, English is well known for its exceptions. In spelling, we put "i" before "e" except after "c" and when sounding like "a" as in "neighbor" and "weigh." We form most plural nouns by adding "s" to the singular, with exceptions like the formation of mice and men. And in interpreting semantics, we try to be sensitive to such subtle distinctions as

> Can I go to Hawaii?
>   *vs.*
> May I go to Hawaii?

or

> I ran into the house
>   *vs.*
> I ran in the house.

But how serious are mistakes in English? If you write "nieghbor" instead of "neighbor," or "mouses" instead of "mice," I know what you mean. I may respond to your Hawaii request with a sarcastic remark about the difference between "can" and "may," but I will know what you're really asking. Likewise, whether or not you distinguish properly between "in" and "into," I can usually tell from the context whether you entered the house or were exercising.

English is, I suppose, as error-prone as languages like PL/I. But English is also much more forgiving. When we converse in PL/I or other programming languages, one error is too many. But when we converse in English, we don't have to be perfect because English has lots of redundant information and we're sophisticated information processors. Mistakes like those above would be fatal only in the strictest of grammar schools.

In stark contrast to PL/I stands Pascal, which was designed in 1968 by the Swiss computer scientist Niklaus Wirth. Pascal is much smaller, much simpler, much easier to learn, much easier to use, and much less error-prone than PL/I. These differences reflect a difference in orientation. PL/I is oriented toward all programs that one might want to write, whereas Pascal is oriented toward the process of programming. PL/I tries to include as much as possible, whereas Pascal tries to exclude as much as possible. Pascal addresses directly the intellectual difficulties of writing and reading programs. Although it was designed originally for the purpose of teaching programming, its general popularity attests to the growing realization that reliable software depends on clarity and simplicity.

The importance of clarity and simplicity in language is hardly a new thought:

> In language, clearness is everything.
>
> —Confucius, *Analects*

and

> Beauty of style and harmony and grace and good rhythm depend on simplicity.
>
> —Plato, *The Republic*

But the stakes in the computer age are even higher than before. If our languages don't encourage us to write clearly, our programs will suffer, and so will we. In the words of British computer scientist C. A. R. Hoare:

The unavoidable price of reliability is simplicity.

The Issue of Style

A good writer writes clearly and persuasively. This is true whether the writer wants the reader to ax a political opponent or fall in love. It's also true when the writer wants the reader to believe that a computer program functions correctly.

To write persuasively, it's important to master spelling, syntax, and semantics. But correct spelling, syntax, and semantics are insufficient for persuasive memoranda and persuasive love letters. They're also insufficient for persuasive computer programs. The meanings of memoranda, love letters, or computer programs are

---

Omit needless words.

Be clear.

Avoid fancy words.

Use definite, specific, concrete language.

Enclose parenthetic expressions between commas.

Avoid a succession of loose sentences.

Express coordinate ideas in similar form.

Keep related words together.

Choose a suitable design and hold to it.

Revise and rewrite.

Figure 10. Advice from Strunk and White.

always expressed by their written texts, but it is the manner in which they're written—i.e., their style—that determines their readability and their persuasiveness.

Clear and compelling memoranda are regrettably scarce in Washington, D.C. The same is probably true—just as regrettably—of love letters, and it's also true of computer programs. There is a simple reason for all three scarcities: Writing well is hard.

It would be nice if we could depend on our language to guarantee good writing, but we can't. For example, French is said to be the best language for love letters. Perhaps this is true, but it doesn't stop English lovers. There are good English love letters and bad French love letters, just as there are good PL/I programs and bad Pascal programs. A language can facilitate and encourage good writing, but it can't offer any guarantees.

Indeed, there are no guarantees. There are, however, some good guides. My own favorite is William Strunk, Jr., and E. B. White's famous little book *The Elements of Style*, in which Strunk and White preach and practice such pithy commandments as those in Figure 10. The commandments come complete with helpful expla-

---

Avoid temporary variables.

Write clearly—don't be too clever.

Choose variable names that won't be confused.

Say what you mean, simply and directly.

Parenthesize to avoid ambiguity.

Use the fundamental control flow constructs.

Avoid unnecessary branches.

Follow each decision as closely as possible with its associated action.

Make your programs read from top to bottom.

Don't stop with your first draft.

Figure 11. Advice from Kernighan and Plauger.

nations and discussions, as well as examples of sins and redemption.

Just as there are guides for writing clear and compelling English, so are there guides for writing clear and compelling computer programs. A good one is Brian W. Kernighan and J. P. Plauger's book *The Elements of Programming Style*, which is similar to *The Elements of Style* not just in title, but also in form and intent. Kernighan and Plauger offer such pithy commandments as those in Figure 11. And, like the commandments in Strunk and White, these too come complete with helpful explanations and discussions, as well as examples of programming sins and their redemption.

A language is a tool for expression, but even a good tool can be used badly. Well-written prose depends on the writer. And, no matter how well the designers of a programming language do their job, well-written programs depend on the programmer. Consider one last commandment:

Do not take shortcuts at the cost of clarity.

Did it come from Strunk and White or from Kernighan and Plauger? You decide.

What Is Structured Programming?

Human nature is attracted to simple rules. Get rich quick. Ten days to a larger vocabulary. Vote Democratic.

So it is with programming. When it became clear in the 1960s that profligate use of the **goto** leads to poorly structured, hard-to-understand programs, a common reaction was to eliminate the **goto**— no more, and no less. In some circles the term "structured programming" came to mean the use of such alternative control structures as **if-then-else** and **while**. Today I still hear the term used in this simplistic way. It's too bad—as a label applied merely because a program eschews the **goto**, "Structured Program" is about as meaningful as "All Natural." Indeed, eliminating **goto**s from your program is no more a sure-fire prescription for achieving reliability than eliminating chemicals from your diet is a sure-fire prescription for achieving health. In programming as in health, it's the overall diet that counts.

The term "structured programming" was coined by Edsger Dijkstra in a seminal monograph called "Notes on Structured Programming." First presented at a 1969 NATO Science Committee conference, the monograph was published in 1972 in the book *Structured Programming*, with accompanying monographs by C. A. R. Hoare and Ole-Johan Dahl. This book is responsible for a revolution in programming. In 1974 Donald E. Knuth—an American computer scientist who, through his popular textbooks and other work, has become one of the most influential programmers of our time—wrote:

> It is impossible to read the recent book *Structured Programming* without having it change your life.

Structured Programming is not a book-length exposition of the advantages of **if-then-else** and **while** over **goto**. In fact, in over two hundred pages, the **goto** isn't even mentioned once—a nice irony. Rather, the book concerns the entire intellectual process of program construction, and it sets forth a new approach. This approach begins with the identification of appropriate abstractions and proceeds with their systematic refinement in a manner that is persuasive about the correctness of the resulting program. I gave you a small glimpse of this process when I developed the algorithm for Aunt Martl's Sachertorte.

Structured programming is not merely a treatment for prevalent **goto**s and other symptoms of unreliable software. Rather it addresses the cause—how we construct programs and how we think about programming. C. A. R. Hoare once described structured programming as

> the systematic use of abstraction to control a mass of detail,

and Niklaus Wirth explained it in these terms:

> It is the expression of the conviction that the programmers' knowledge must not consist of a bag of tricks and trade secrets, but of a general intellectual ability to tackle problems systematically. . . . At

its heart lies an *attitude* rather than a recipe: the admission of the limitations of our minds.

The development of structured programming did lead to new patterns of language usage—decreased use of the **goto** being one example—but these new patterns resulted from thinking about programming in new ways, not the other way around. One new way of thinking came from the recognition of programming as a literary activity. But just as important was the recognition that programming is also a mathematical and an architectural activity— aspects of programming to which I will now turn. Before I do, however, let me exploit a remark of George Santayana, who once wrote:

To turn events into ideas is the function of literature.

To turn ideas into events is the function of programming.

CHAPTER 10

Programming
As Mathematics,
Programming
As Architecture

The literary view of programming emerged as a response to the software crisis; a good programming style makes programs easier to read, and readability leads to reliability. But readability isn't enough. If we are to argue about a program's correctness, for example, we have to know what correctness means. And while we want programs to be clear and correct, we also want programs that are easy to change; unfortunately, ease of change does not follow necessarily from clarity and correctness.

Mathematics and engineering have shaped the modern view of programming in ways that address these and related issues. The resulting view is intellectually attractive and interesting, and it would be so even if it weren't important. But it *is* important. Learning something about it will give you a sense for current attempts to cope with the software crisis. And if you write programs yourself, or contemplate doing so, some insight into the modern view can also help you to write better programs.

When people hear about the role of mathematics and engineering, they often assume that programming has been transformed from an art into a science. Not true. Mathematics and engineering have brought discipline to programming, but they have also brought their own artistic sides. In thinking about literature, mathematics, and architecture, we appreciate such aesthetic qualities as beauty, style, and grace. With the modern viewpoint, these same qualities have come to programming. Indeed, programming is unique as a human activity that combines the challenges, skills, and pleasures of artistic creation, mathematical insight, and engineering achievement.

Mathematical Meaning

Computer programming and mathematics have always been intimate. Indeed, the first computers were designed and built to serve as tools for mathematics—in the late 1940s, computers like ENIAC tabulated mathematical functions, computed ballistic trajectories, and analyzed the predictions of atomic physics (see Chapter 1, Figure 1). Throughout these applications, however, the relationship between programming and mathematics was one-sided. Programming existed almost exclusively to assist in mathematical studies. But, except for studies about the numerical accuracy of programs and studies about the theoretical possibility of solving certain problems by means of computer programs, mathematics wasn't used to study programming itself. Mathematics was the subject of programming, but it took another twenty years before programming became a subject of mathematics.

Where's the Math?

One of the reasons it took so long is that the role of mathematics isn't obvious. It's obvious that programs can compute numbers and therefore that programming can serve mathematics. But how can mathematics serve programming?

The answer is found in the inadequacy of program testing. We know that testing can't establish correctness—we can't enumerate and check the results of every possible combination of program inputs. Rather, using arguments based on the program's written text, we must demonstrate—without trying every case—that the program works in every case. Stated differently, we must reason about the meaning of the program as an abstraction (there's that word again), conceived apart from the special cases of particular input values. Such abstract reasoning is precisely what we use mathematics for; e.g., we can demonstrate explicitly that the sum of four and six is an even number, but we can also use mathematics to prove in abstract terms that the sum of any two even numbers is an even number. It follows that mathematics might be useful in reasoning about the meaning of computer programs in order to prove whether or not they are correct.

What Is a Proof?

As a boy, when I was confronted with a statement I couldn't or didn't want to believe, my reaction was simple. I would say, emphatically and in a whining voice, "Oh yeah? Prove it!" Now that I'm older I try not to whine.

Asking for proof is part of the truth seeking process. We do it naturally and in many different contexts—in stores, in courtrooms, and in math classes. Although we like to think of proofs as establishing certainty, they don't. Receipts can be forged, innocent people sometimes are convicted, and mathematicians make mistakes. Proofs, like other human activities, are fallible.

If a proof doesn't always establish truth, it does establish a convincing argument. Whether the proposition in question is

The sum of two even numbers is an even number,

or

The Earth goes round the Sun,

or

The patient is dead,

or

The defendant is guilty,

or

The program finds which employee has the largest salary,

a proof is a convincing presentation of evidence in favor of the proposition. When we say that we've proven something beyond any doubt, we really mean that we've reduced our doubt to some acceptable level. This fact is recognized in our legal system—we don't ask for absolute proof, we ask for proof "beyond a reasonable doubt."

Proving Program Correctness

Our legal system also presumes that people are innocent until proven guilty. Computer programs, however, deserve harsher treatment—we should presume that programs are guilty of erroneous behavior until proven innocent. It would be nice if we could remove particular computer programs from suspicion by proving absolutely that they behave correctly. But we can't. Proofs of program correctness are as fallible as other kinds of proofs. In programming as in politics, some doubt always remains.

The problem is to reduce our doubt to an acceptable level. Testing can help a little, but it's an inadequate solution. No matter how often we execute a program, substantial doubt will remain unless we have other evidence for its correct behavior. (The situation in politics is somewhat similar: execution can't remove doubts about a citizen, although it can remove the citizen.) That other evidence must come from studying the program's written text.

To evaluate the correctness of a program is conceptually simple: you start with a statement of what the program is required to do, and you read the program to determine whether or not it meets the requirements. But even the starting point is difficult to achieve; it's hard to write a complete and unambiguous statement of requirements. And it's just as hard to determine whether or not those requirements are met. When we attempt to do so, we usually consider each statement of the program in turn and analyze how our understanding so far is affected by the statement currently under consideration. This procedure leads to two problems: first, chains of reasoning can get long and complicated; and second, success requires that we understand clearly the meaning of every statement that we encounter. This second requirement may not sound like a problem—how hard can it be to understand a single statement?—

but it is a problem. The problem arises from the natural language descriptions that typically are used to define the semantics for each type of programming-language statement. These descriptions are notorious for containing ambiguities and oversights that leave the statements open to slightly different interpretations by different people. For example, the interpretation by someone who writes programs in the language might differ slightly from the interpretation by someone who writes a compiler or interpreter for the language.

Mathematics can help with all of these problems. Perhaps you think of formal mathematical notation as an impenetrable forest of foreign letters, squiggly lines, and esoteric-looking symbols—which is how I happen to think of anything written in Chinese, Sanskrit, or Egyptian hieroglyphics—but, whether or not you understand the notation, it's important to understand why mathematical notation has proved to be so useful. In particular, the notation is useful in writing precise and unambiguous descriptions, which is exactly what's needed for specifying software requirements. Moreover, the notation makes it easier to define precisely and unambiguously the meaning of individual statements in a complicated chain of reasoning, and easier to chain the steps together, two advantages that are exactly what's needed when we try to understand the meaning of a program. To get these advantages, we need only use mathematics instead of natural language to define the meaning of each programming language statement.

This is exactly what the American mathematician and computer scientist Robert W. Floyd did in a 1967 paper entitled "Assigning Meanings to Programs." Here's a small example that illustrates Floyd's basic idea: Suppose you're reading a Pascal program and come to the statement

```
C := A + B;
```

and suppose you're convinced that, at this point, the value of A is six, and the value of B is at least four. It follows that, once the above statement has been executed, the value of C will be at least ten. To a limited extent, that's all there is to it.

Floyd suggested that the meaning of a program be defined in

terms of mathematical propositions about the values of program variables—propositions like $A = 6$, $B \geq 4$, and $C \geq 10$ (read "A equals six," "B is greater than or equal to four," and "C is greater than or equal to ten"). Moreover, he suggested that the meaning of any particular statement in the program be defined by rules that transform propositions that are known to be true before the statement is executed ($A = 6$, $B \geq 4$) into propositions that will be true after the statement is executed ($A = 6$, $B \geq 4$, $C \geq 10$). This approach has two fundamental virtues. The first is that the transformation rules for all of the statement types in a programming language together give a mathematical definition for the semantics of the programming language. The second is that, given a particular program in that language, and given a true proposition about the values of the program's inputs, the transformation rules make it possible to consider the program's statements one at a time and derive—at least in principle—the consequences of the entire program. The results can then be compared with the program's specification in order to determine whether or not the program is correct.

Together with related work by C. A. R. Hoare, this work by Floyd stimulated a whole new way of thinking about programming and programming languages. The semantics of programming languages and programs could now be studied mathematically. It became possible not only to write and read computer programs, but also to prove their correctness.

Who Proves Program Correctness?

Hardly anyone. This may relieve you, especially if you are contemplating writing programs yourself, but it may also seem surprising in light of the glowing reverence with which I described the role of mathematics. Let me explain. There are three principal reasons for the limited extent of program-proving activity today. First, it's tedious to derive a proposition that expresses the meaning of any program that isn't small and simple—the mathematical expressions quickly become unmanageably large and complicated. Another reason is the limited availability of formal semantics— those mathematical rules that determine how propositions about program variables are affected by the execution of programming-language statements. Formal semantics have not been defined for large portions of the programming languages that are in widespread use, primarily because significant technical problems arise during attempts to do so. Finally, even when it's feasible, proving correctness is hard—mathematics is not an easy subject.

Program-proving today may be cumbersome, impractical, and hard, but its effects on programming have been profound. A principal reason is that formal methods of program-proving can also be applied informally. If you understand the ideas and methods behind formal proofs of correctness, it's easier to write correct programs and it's easier to argue convincingly about the correctness of programs you read and write.

It's no accident that the mathematical viewpoint of programming developed during the same period as the literary viewpoint I discussed in the previous chapter. The two viewpoints face different sides of the same coin—a coin that buys clarity and ease of understanding. Well-structured, well-written programs are more amenable to formal or informal correctness proofs than are badly structured, poorly written programs. For example, the fewer **goto**s there are in a program, the easier it is to derive a mathematical statement of the program's semantics. Furthermore, abstractions useful in the development of a structured program also tend to be useful in proving the program's correctness. Indeed, structured programming methods and simpler programming languages were developed in large part to facilitate correctness proofs. These advances permit steps in refining a program to be taken side by side with steps in proving—formally or informally—the program's correctness. Donald Knuth has described the situation well:

Fifteen years ago computer programming was so badly understood that hardly anyone even *thought* about proving programs correct; we just fiddled with a program until we "knew" it worked. At that time we didn't even know how to express the *concept* that a program was correct, in any rigorous way. It is only in recent years that we have been learning about the process of abstraction by which programs are written and understood; and this new knowledge about programming is currently producing great payoffs in practice, even though few programs are actually proved correct with complete rigor, since we are beginning to understand the principles of program structure. The point is that when we write programs today, we know that we could in principle construct formal proofs of their correctness if we really wanted to, now that we understand how such proofs are formulated. This scientific basis is resulting in programs that are significantly more reliable than those we wrote in former days when intuition was the only basis of correctness.

Unfortunately, despite its importance, the mathematical viewpoint of programming is not widely known. For example, Knuth's observations are reasonable and accurate, but they were written ten years ago! Nevertheless, introductory programming courses routinely are taught today without even mentioning that the mathematical viewpoint exists. This year perhaps a million people will write their first program—probably a BASIC program for a small computer from Apple, IBM, Radio Shack, or Commodore—without anyone explaining to them that there's more to program correctness than intuitive fiddling. They will think of themselves as joining the mainstream of the computer age, when in fact they are being introduced to programming as it was understood twenty-five years ago.

Software Buildings

Designing and building a house is an exercise in stepwise refinement. At the start, there's just an abstraction: the idea of a house. In a series of familiar steps, architects and builders refine that abstraction until they produce a special case: a physical structure in which you or I might live. The steps of refinement involve settling on location, number of rooms, overall size, actual layout, detailed plans, construction materials, appliances, lighting, and so on. Although the analogy can't be pushed too far, the overall process is somewhat like designing and building a computer program. At the start, there's just an abstraction: the idea of a program to solve a particular problem. Programmers refine this abstraction in a series of steps that involve settling on a detailed specification, a general algorithm, a specific program, a computer, a compiler, an assembler, and so on.

Architects and programmers have many concerns in common. Whether you propose to construct a building or a program, your concerns include questions such as these:

What requirements must it satisfy?

How big will it be?

How long will it take to build?

How much will it cost to build?

What resources will be required to use it?

How much will it cost to use?

What will be its capacity?

How well will it perform?

Can part of it be used before it's completely ready?

How easy will it be to modify?

How easy will it be to extend?

To take the analogy a step further, remember that in Chapter 7 I described programming as a process of designing and building abstract machines. An architect's plans and blueprints can likewise be thought of as the design for an abstract machine. These products stop short of a physical building and therefore describe an abstraction—numerous special cases can be built, which is exactly what happens in the modern housing development. Furthermore, although we don't usually think of buildings as machines, machines they are, with both a static structure and a dynamic structure—doors and windows open and close; elevators move up and down;

water, electricity, and heat flow. As the architect Le Corbusier put it:

A house is a machine for living in.

Information Structures

The words "instruction" and "structure" come from the same Latin root: the verb *struere*—to build. The structural nature of programming is further emphasized by modern terminology: structured programming is the disciplined construction of programs out of suitable data structures and control structures. I like these terms. They emphasize not only that programs are structures, but also that structure is the key to coping with software's intrinsic complexity.

Like buildings, programs have a static structure and a dynamic structure. The static structure of a program is evident from its written text. The static structure includes a description of the program's data structures, which are chosen for convenience in storing and manipulating the information processed by the program, and the program's control structures, which determine the flow of control.

As for the program's dynamic structure, like the dynamic structure of a building, it evolves in time. The program's control structures operate to determine the flow of information through the program's data structures, somewhat like valves and switches determine the flow of hot water in an electric hot-water-heating system. In a program, of course, information is the key commodity, and it is information-processing requirements and methods that determine a program's overall structure.

Actually, information-processing requirements and methods also affect the structure of buildings. An obvious example is the modern skyscraper, bristling with microwave communications links and humming with word processors and electronic mail. In fact, information-processing requirements and methods are partially responsible for the structure of all skyscrapers, not just modern ones. Although we tend to think of the skyscraper as having been made possible by the invention of the elevator and various relevant construction techniques, the skyscraper's economic viability depended greatly on an information-oriented invention—the telephone. To distribute and exchange information before the telephone was invented, businesses relied heavily on human messengers. If the same were true in skyscrapers, many more elevators would be required. Elevators take up a significant volume as it is; if they took up much more, the remaining office space wouldn't be worth the cost. Take away the telephone, and you would take away the characteristic skylines of our cities.

The overall function of a building is served more by its static structure than by its dynamic structure. For programs, it's the other way around: A large program might comprise a million machine-language instructions—the basic building blocks of software—but it can take several hours of execution for that program to perform its intended function. In two hours on a computer that executes a million instructions per second (fast, but still modest by supercomputer standards), about seven billion instructions get executed. Constructed from billions of their basic building blocks, computer programs are elaborate structures by any measure. For comparison, imagine that seven billion pieces are used to construct a skyscraper 200 feet square and 1500 feet tall—that's a total volume of sixty million cubic feet, considerably larger than the Empire State Building. If this volume were divided into seven billion pieces, the average piece would be a cube with sides about 2.5 inches. In fact, skyscrapers are constructed out of much larger pieces. Computer programs are mankind's most elaborate artifacts.

The Water Heater Module

While standing recently in a cold shower, I had occasion to think about replacing my hot water heater. Because I had been thinking about software engineering before the water turned cold, I was impressed by the extent to which I could choose a new water heater without thinking about air conditioners, room heaters, telephones, windows, or practically anything else except the number of people in the house and the number of dollars in my bank account.

It's simple to select a new water heater because most water heaters conform by convention or law to a variety of standards—

examples include standard shapes, standard-size pipe fittings, standard safety features, standard units for describing energy consumption, and standard units for describing capacity. Such standards make life easier for the architect, the builder, and me; I could order a 50-gallon, gas-fired hot water heater and be done with it. Without the standards, my water heater problem would have been harder to solve, but it would still be manageable because I would have had to consider only the externally visible characteristics of water heaters: their size; their shape; their needs for electricity, gas, oil, or other energy resources; the amount of hot water they produce; the positions and dimensions of their connections; the convenience and safety of their controls; statistics about their reliability; and so on. To use a technical term, these characteristics define the interface of a water heater.

The point is that, in selecting water heaters, we need only consider their needs and their products, not how their insides work. In a gas-fired water heater, for example, such details as the size and shape of the burners, the method for lighting the burners, and the method of measuring temperature are all—figuratively speaking—secrets of the water heater design, hidden behind the interface. In some cases the details are literally secret—trade secrets protected by law.

Such secrecy has enormous practical advantages. To begin with, it restricts the number of people who are forced to worry about the internal details of water heaters. If you're in the business of designing water heaters, then these details are properly your concern, but the rest of us don't have to clutter our brains with them when we think about designing and building houses. Moreover, because we don't consider them when we design and build houses, these details are free to change later when we replace one water heater with another.

How different home life might be if builders routinely made assumptions that violated the conventional water heater's interface. For example, suppose some earlier residents of my house had rigged up a nonstandard energy-saving device that closed the flue whenever the burners were off. (The flue, which usually goes up a chimney, is a tube that carries off the hot gases that arise when the water heater's burner is on.) Suppose that the device was controlled by a temperature-sensing probe installed right over the burner—when the burner came on, the flames heated the probe, and the device opened the flue whenever the measured temperatures reached some threshold. If I then came along and switched water heaters, all sorts of problems could have occurred.

For one thing, I might have overlooked the device, in which case I might have connected my new water heater to a permanently closed flue. If I did notice the device and tried to make it operate with the new water heater, I might have had to deal with all sorts of differences between the old and new heaters, differences that usually don't matter. For example, the temperature probe might not easily fit the new burner. If I managed to install the probe anyway, the device still might not work properly—some new water heaters could have a cooler burner, cool enough never to trigger the device. Other water heaters could have a hotter burner, hot enough to melt the probe before the device opened the flue. These aren't the only possibilities; I could think of several more while watching my house burn down.

The hypothetical energy-saving device causes problems because its installation and correct operation depend on special assumptions about burner size, shape, position, and temperature—assumptions that aren't valid for all water heaters. Stated differently, the device would have introduced some undocumented and nonstandard assumptions into the interface of my water heater. Fortunately, such a dependence on undocumented assumptions doesn't happen often with water heaters; unfortunately, it happens all the time with software.

The Secret of Information Hiding

Buildings are complicated machines. To design and build them effectively, we've learned to cope with their complexity by breaking them into manageable units; call them subsystems, modules, abstractions (e.g., hot water heaters). For each subsystem we identify the characteristics that will affect the design of other subsystems (e.g., electrical load, pipe diameter), and we require that these characteristics satisfy stated specifications. In other words, we define the interfaces between subsystems. Then, provided that our results meet the interface requirements, we are free to design the subsystems independently of each other. When we solve problems in this general way, we exploit one of the few principles that applies equally well to mathematics, engineering, politics, and warfare: divide and conquer.

The principle also applies to software. The resulting units of independent development are commonly called **modules**, and the systematic development of software structures using a manageable set of modules is called **modular programming.** Modular programming was recognized early as a desirable method of software development—the term even predates the earliest work on structured programming—but desire is one thing, fulfilment another. That the modular approach could be effective was obvious from other engineering fields, but how to make it effective in software development wasn't obvious at all.

The difficulties inherent in modular programming arise from the invisible interfaces and the curse of flexibility that characterize the software medium. It's worth recalling the main points from Chapter 8: When designing physical objects like buildings, airplanes, and cars, the physical separation of various functions guides an effective decomposition into modules. Moreover, the obvious cost of physical connections between modules helps to keep interfaces simple. In contrast, software has no physical structure to guide the decomposition into modules, and—exempt from the discouraging influence of physical constraints—the software medium encourages complicated interfaces. Furthermore, along with the complexity come plenty of undocumented, subtle assumptions, like the one that was responsible for the space shuttle bug. As a result, it's hard to write large bug-free programs, and it's hard to modify them without introducing new bugs. After a series of modifications are made without understanding their full effects on the overall structure, the program becomes unstable, like a house of cards.

A key to making modularity work in buildings, airplanes, and cars is the accurate specification of the interfaces between modules; here, I'm using the term "specification" in the engineering sense of a statement of requirements. When we build water heaters, we require that they meet certain specifications, which is why I knew that a new water heater would connect easily to the plumbing system in my house. Stated differently, I was counting on the water heater satisfying a standard interface specification.

In retrospect, it seems obvious that the accurate specification of module interfaces is a requirement for the construction of reliable software. But the connection was overlooked until 1972, when the American computer scientist David L. Parnas introduced a method of documenting module interfaces with **abstract specifications**— specifications that constrain the externally visible behavior of software modules while leaving the details of internal operation to be chosen by those who write the actual programs. Precise specifications were also addressed in the earlier work on program-proving by Floyd and Hoare—you can't prove that a program meets its specification unless you have the specification—but specifications were addressed more as a side issue. The emphasis was on verification rather than specification, and abstract specifications were not considered at all. We now realize that precise, unambiguous, abstract specifications are one of the keys to software reliability. Indeed, the study of specification methods is today one of the most important and active areas of software-engineering research.

Unfortunately, before you can write abstract specifications for a set of modules, you have to choose the modules. The problem is analogous to this one: Having never seen a house, you're given a description of the functions served by houses—the human needs that houses must meet. From this description you have to decompose the problem of house design into a set of smaller problems with clearly defined interfaces. Your job is to come up with the likes of bedrooms, bathrooms, hallways, doors, windows, lights, water heaters, and other useful abstractions. Simply put, the job in decomposing software is the same: finding useful abstractions— the water heaters of software.

Constructing reliable, modifiable software therefore requires the solution of two problems: first, choosing the right set of modules; and, second, documenting them with abstract specifications that constrain their interfaces. These problems are clearly related, and it isn't surprising that, while finding a solution for the second one, Parnas also found a solution for the first. To make it easier to appreciate his solution, let me return once again to the water-heater module.

I mentioned earlier that various internal details of water heaters can be thought of as secrets, hidden behind the water heater's interface, and that this secrecy makes it easier to design houses, to design water heaters, and to replace water heaters. The hypothetical energy-saving device was troublesome because its installation and correct operation depended upon assumptions that were not part of the standard water-heater interface. Responsible building contractors try to avoid such trouble by sticking to the standard interface, which they can do provided that the interface has been designed well—i.e., provided that the interface hides those aspects of water heaters that may change from water heater to water heater.

From this perspective, Parnas's solution for software is equally clear: a module's interface should shield other modules from internal details that may change later. Moreover, given information about the ways in which a software system may change after its initial development, a good decomposition into modules is one that confines each likely change to a single module. Because detailed information about the operation of such modules is hidden behind their interfaces, they are called **information-hiding modules.** Their secrets are those aspects of the overall system that are subject to change.

Messing with Naval Messages

Perhaps a real-life example will make it easier to understand and appreciate the importance of information-hiding modules. The example concerns one of the Navy's message-processing systems. These computer-based systems are sort of seagoing Western Unions, and they do what you might expect—they assist in writing, transmitting, receiving, and delivering formal messages that are similar to conventional telegrams.

A few years ago Parnas studied the software structure of one of these message-processing systems to see how it was divided into modules. I've mentioned several times that physical structure is a useful guide to the modular decomposition of physical systems. In software the closest thing to such a physical structure is the distinction between various storage devices—for example, main memory, disks, and tapes—and it's a natural tendency for people to use these distinctions as a guide to selecting module boundaries. Because these distinctions are not sufficiently elaborate to yield many modules, people also turn to the dynamic structure of the tasks they are seeking to automate; temporal proximity replaces physical proximity as a criterion for assigning components to the same module. Parnas found exactly these criteria apparent in the message-processing system he studied—there were modules for creating messages, storing messages on disks, retrieving messages from disks, transmitting messages, and receiving messages.

This decomposition into modules seemed sensible to the system's designers, and it appeared to be adequate in practice until the Navy decided to change the format of their messages. Consider the effect of a change that requires the destination portion of messages to be enlarged—a change just like the change from five- to nine-digit ZIP codes that's currently being instituted in the U.S. Mail system. Unfortunately, the message format was the worst kept secret in the system. Messages were passed from module to module essentially in raw form, with a fixed number of characters devoted to each portion of a message, so that changes in the format of messages would require changes to every program that manipulates messages. Every module would have to change—a laborious and error-prone prospect.

Consider, by way of contrast, a design in which the message format is the secret of a single module. Other modules obtain the different portions of messages from this module without knowing the actual format of the message. For example, to obtain the destination portion, they might invoke a program called **DESTINA-TION**. A change in the format of messages could require that the **DESTINATION** program be rewritten. But other modules would still obtain message addresses by invoking the program called **DES-TINATION**, and therefore would require no changes at all.

This example is no exception. Some time later Parnas, others, and I based a hypothetical message-processing system on this real example, and we used the system as an illustration in a course on software engineering. Six people came to the course from the facility where Parnas had studied the original message-processing system. During the course these six people all claimed to recognize the "real" system behind the example, but they each named a different system!

Like so many other concepts that are useful in software design, information-hiding modules are abstractions; their externally visible behavior is conceived apart from special cases that implement the required behavior in different ways. Abstract specifications describe the externally visible behavior and hide the actual implementation. To paraphrase something that Parnas said to me, information hiding and abstraction are just two sides of the same coin. One side concerns what is hidden. The other side concerns what is seen.

Modular programming is an iterative process. The overall system is first decomposed into a manageable number of modules (seven or so), each documented by abstract specifications and each hiding a big secret. The modules are then decomposed in turn into sets of submodules that hide smaller secrets. This process continues until the final modules can each be implemented by a small program written by one person. The resulting system is not only more likely to withstand changes, it's more likely to be correct in the first place. As Parnas put it,

Constructing big programs without ever writing a big program is the key to constructing correct big programs.

The Role of Language

Whether you construct buildings or programs, the quality of your product depends on the quality of your tools. In the case of program construction, the basic tools are programming languages. We've learned that the safe construction of mankind's most elaborate artifacts can best be accomplished by assembling hierarchical layers of abstraction—from machine language to assembly language to high-level programming languages to data abstractions and information-hiding modules. In this context, the most important role of a programming language is as a tool-building tool—abstractions built on one level are the tools for building the next level of abstraction.

Suppose "management" asks you to write a program that maintains information on your organization's employees and computes various statistics about them. Let's say that you need only consider their salaries, their Social Security numbers, and their telephone numbers. If you write a machine-language program, you might allocate three memory locations to each employee, with each location storing one of these numbers. If you write an assembly-language program, you might make these same decisions, but the program would be much easier to write because you wouldn't have to worry about which specific memory locations are used for each particular employee. If you write a FORTRAN program, for each employee you could create a little data structure containing three integers. As in the assembly-language program, you would be relatively free from concern about the correspondence between employee data and memory locations. Moreover, in many respects you could manipulate each employee's data as a single unit, which makes it easier to write the program and easier to modify the program later if you were required to keep track of more than three numbers for each employee.

If you write a Pascal program instead of a FORTRAN program, you could also create a little data structure for each employee, but the possibilities are more general since the data structure need not be composed of three integers. You could, for example, use an integer for salary and strings of characters for the Social Security number and the telephone number. This would allow you to put hyphens into telephone numbers and Social Security numbers (where they "belong"), and would enable the Pascal compiler to stop you from performing meaningless operations like adding a telephone number to a salary.

It's much easier to write the Pascal program because the data structures you create allow you to think and write in terms of a meaningful abstraction—employee records. Generally speaking, Pascal helps because it has some general features for defining new abstractions and using them as building blocks in programs, as do a variety of other popular languages, including Logo and Lisp. In languages like FORTRAN and BASIC, however, the basic building blocks essentially are predetermined—with FORTRAN, it's possible to build with abstractions in a limited way, but you have to work at it; with BASIC, it's almost impossible. Unfortunately, because inexpensive implementations of BASIC exist on practically every small computer, BASIC is the language most often used to introduce computer programming. To give people their formative programming experiences with a language that lacks the single most important concept of modern programming—abstraction—is ridiculous.

Providing good tools for building hierarchical abstractions is today one of the most important areas of programming-language design. Many languages of recent design reflect this emphasis, and they provide facilities for abstraction that go far beyond those of Pascal. Most of these languages exist only as research tools or as languages used for the development of commercial software. A few of them may eventually become generally available—two examples are Smalltalk, which was designed by the Software Concepts Group at Xerox's Palo Alto Research Center, and Modula, which was designed by Niklaus Wirth (the designer of Pascal). The U.S. Department of Defense's new programming language, ADA, designed by the French computer scientist Jean Ichbiah and others, also provides extensive facilities for abstraction; it, too, could become successful and generally available, although its large size may prove to be too cumbersome. A noteworthy ADA feature is an abstraction mechanism called "packages." Packages were inspired by Parnas's work, and they provide an explicit mechanism for using a form of information-hiding modules to build large software systems.

Over the last fifteen years, it has become apparent that program construction is as important as program writing. It follows that good programming languages should do more than facilitate the writing of clear and compelling programs; they should also facilitate the construction of large software systems. The key is to allow the programmer to build with abstractions. The right abstraction confers enormous advantages, something that's as true in building construction as it is in program construction. Suppose that you can think and build not just in terms of constant-length steel rods, but also in terms of equilateral triangles and other regular polygons. Suddenly it's easy not only to build scaffolds, but also to build geodesic domes.

The Role of Models

Most people who contemplate building a house begin by looking at other houses. The same is true of airplanes, kites, bridges, and every other engineering product I can think of. There are a variety of reasons for this behavior, ranging from laziness to good sense. Existing engineering products usually represent several generations of incremental discovery. To ignore them would be foolish— few people are capable of getting as far by themselves. Existing models help us to start out further ahead, and to start out quickly. Unfortunately, our economic and intellectual investments impart a historical momentum that makes it hard to deviate from the conventional path. The study of existing models may enhance our capabilities, but so may it narrow our vision.

Software is no exception. When we design programming languages, compilers, and operating systems, we build on the discoveries of our predecessors. The same is true when we design programs for naval message systems, aircraft flight control, and PacMan. Deviating from these models is hard—in this regard, software models are as influential as other engineering models.

One might excuse the lack of variation in house designs by arguing that we've already settled on the best ones. This may even be a valid excuse; perhaps radical inventions like the geodesic dome are infrequent because there are few such inventions left to be made. But a similar excuse for software is invalid. There simply has not been sufficient software-building experience to yield the best designs—on the contrary, as the software crisis testifies. Unfortunately, even when you understand what's wrong with existing designs, and even when you discover better design principles and better design methods, better designs are still hard to come by. You may understand and appreciate modern software-engineering ideas, but that doesn't mean you'll be able to apply them effectively.

A good example arose from the software-engineering course that I mentioned earlier. Others and I taught this course to several hundred key naval personnel during the period 1976–1981, hoping to cause a dramatic change in the Navy's software practices. Many of our students were receptive. They converted to the modern viewpoint, they returned to their jobs with enthusiastic idealism, and they proceeded to do their jobs more or less exactly as they had done them before.

Actually, it wasn't quite as bad as that. When we followed up, we found many instances in which the course had helped. But no drastic changes had occurred. We felt a bit like revolutionaries who, having just pulled off a series of political rallies, were mystified by the government's failure to fall instantly. The problem, we finally admitted, was this: we *told* our students how to build Navy software, but we didn't *show* them how to build Navy software. We did have examples and programming exercises in the course, but we didn't have any real, live examples of Navy software systems that had been constructed the "right" way.

People need those examples. And they need examples of "their kind" of system—those who build message systems need examples of message systems, those who build flight software need examples of flight software, and so on. Once we realized this, we decided to provide an example by having Parnas lead a project to rebuild the obsolescent flight program of a particular naval aircraft. (This is the project that got us into trouble when we wrote that the existing software "is not fully understood by the maintenance personnel.") The project was started in 1977, and it was supposed to last about three years.

As I write this in 1984, the project is still going strong. It's taking so long in part because of a lean budget, but primarily because of an ironic twist. The project's purpose is to provide a new model; it's taking so long because the project team has no model to follow! Although the team has yet to finish the entire model, its early products are already being used as models in the development of some related Navy systems.

In program construction as in building construction, it's hard to break new ground. To do so requires technical expertise, intellectual insight, and creativity. It's hard, but it's important. If modern insights into software development are to improve software reliability, we must provide good models. In the words of Winston Churchill:

We shape our buildings; thereafter, they shape us.

What Is a Programming Language?

Posed after several discussions of programming, this may seem like an odd question. But the question goes to the heart of the changes that have taken place over the past thirty years, and in doing so it provides a nice perspective.

Among the many views of programming languages, two stand out. The first—and it really was the first—sees a programming language as a convenient notation for instructing computers. In this view, the computer hardware is a main focus of attention, and a principal goal of programming is to achieve high execution speed and small program size—two qualities that loosely define **efficiency**. FORTRAN exemplifies this view. A specific example is the so-called "arithmetic-IF," a FORTRAN statement that wraps three conditional **goto**s into one statement. The statement transfers control to one of three destinations, depending on whether a given quantity is less than, equal to, or greater than zero. For example, if I is an integer variable, the FORTRAN statement

IF (I) 23,14,67

essentially is equivalent to the following sequence of three Pascal statements:

if I $<$ 0 **then goto** 23;
if I $=$ 0 **then goto** 14;
if I $>$ 0 **then goto** 67;

It doesn't take too many of these to destroy the readability of a FORTRAN program, and you might wonder what possessed the designers of FORTRAN to include such a thing.

The reason is simple: the designers had in mind the IBM 704 computer, and they included the arithmetic-IF primarily because it would be easy to translate into efficient IBM 704 machine code. In particular, the 704's instruction set included a "Compare Accumulator with Storage" instruction; the effect of this instruction was to subtract the contents of the accumulator register from the contents of any desired main memory location, and then to skip zero, one, or two instructions depending on whether the result was less than, equal to, or greater than zero. The three-way branch could be accomplished by putting two unconditional transfer instructions immediately after the "Compare Accumulator with Storage."

FORTRAN's designers saw FORTRAN as a high-level notation for IBM 704 programs, and their main concern was to translate this notation into efficient machine-language programs. The machine-independent expression of algorithms was practically the farthest thing from their minds. John Backus, who led the FORTRAN project, has written:

> We certainly had no idea that languages almost identical to the one we were working on would be used for more than one IBM computer, not to mention those of other manufacturers.

Today there's a different emphasis, and it's reflected in the second view of programming languages. A program isn't useful if you can't afford to run it, but the software crisis has demonstrated that the goal of efficiency can easily lead us astray from reliability. Efficiency is important, but the efficient execution of incorrect programs isn't particularly useful. We all want a fast car that gets good gas mileage. But we also want a car that's safe to drive and doesn't require constant repairs.

The second view of programming languages sees them not as convenient notations for instructing machines, but as convenient notations for expressing algorithms. The difference may seem subtle, but it's important. The first view focuses on the machine's convenience, while the second view focuses on our convenience. As Dijkstra put it:

> There are two views of programming. In the old view, the purpose of our programs is to instruct our machines; in the new one, it is the purpose of our machines to execute our programs.

Machines only have to execute programs, but people have to read programs and judge their correctness; notations that are convenient for instructing computers are often inconvenient for expressing algorithms and reasoning clearly about their correctness.

The software crisis resulted primarily from our failure to manage complexity. Natural constraints bring discipline to the design of physical objects, thereby helping to control their complexity, but nature doesn't help with software. If we are to control software complexity, then we must impose discipline ourselves. Simpler programming languages, attention to programming style, structured programming methods, proofs of correctness, information-hiding modules—all of these are tools that contribute to such a discipline. They do so by helping us to write programs and by helping us to think. They are antidotes for the curse of flexibility. It's not just that the new ways are better—the old ways are simply inadequate.

CHAPTER 11

Electronic Cretins

It has long amazed me that some people can laugh at the question

Can computers love?

while they take seriously the question

Can computers think?

To the first question they respond, "Ah, but what is love?" To the second they respond with impassioned arguments, perhaps pro, perhaps con. I don't get it. It's reasonable to ask "How do computers behave?" and it's reasonable to ask "Can a computer appear to be thinking?" But the question "Can computers think?" seems meaningless to me. Moreover, however we describe that elusive quality or process we call "thinking," how can we be sure that people really work that way? Or, as B. F. Skinner wryly put it:

The real problem is not whether machines think but whether people do.

Besides, as Edsger Dijkstra is fond of pointing out, if you look in Webster's *New Collegiate Dictionary* you find

intelligent, adj. . . . 3: able to perform some of the functions of a machine.

Can computers think? Ah, but what is thinking?

As meaningless as it may be to ask whether computers can think, the question does raise meaningful issues. The computer is the most general, capable, adaptable tool that has ever been invented, and it's natural to ask about its limits. People know that computers can serve as word processors and tax calculators, and they ask "What else can it do?" They worry that computers may someday be more capable than they are. Such worries are not new. In May 1963, after astronaut Gordon Cooper had flown a manual reentry in his Project Mercury space capsule, President John F. Kennedy welcomed him to the White House with these words:

> I think that one of the things which warmed us most during this flight was the realization that however extraordinary computers may be, we are still ahead of them, and that man is still the most extraordinary computer of all.

Computers vs. Brains

Many people think of computers as "electronic brains." It's a natural enough analogy, mainly because both brains and computers are centers for information processing. Your brain receives, stores, and processes information, dispenses results, and controls your biological equipment. When properly programmed, computers can do likewise, except that they control electromechanical rather than biological equipment. Beyond these functional similarities, however, computers and brains have virtually nothing in common.

To begin with, the electronic circuits in a computer are not analogous to brain cells. The two differ in appearance, in structure, and in principles of operation. The key functions of information storage and information processing are served in computers by physically different components. In a typical computer, one finds separate CPU and memory units; but even in computer designs where processing circuits are intermixed with memory circuits, the two functions remain distinct. In the brain they are not distinct; they're distributed throughout the brain and intermixed in ways that we don't understand. As for the mechanisms by which computers and brains work, electromagnetic activity is clearly important to both. But in computers electromagnetic activity is the primary agent of information storage and processing, while in the brain, biochemical processes also play fundamental roles. Although much is known about the physiology of the brain, we really don't understand how it processes information, how it controls the body, how it thinks.

Although computers are internally unlike brains, their external performance is another issue. Some things are hard for us but easy for computers. Other things are easy for us but hard for computers.

Counting

One of the most obvious differences between computers and brains is in the speed of simple arithmetic operations. Lots of computers can add a million numbers in a second. How many numbers can you add in a second? In a sense, the ability to solve numerical problems quickly is the fundamental advantage of computers. Indeed, the first ones were built to replace and surpass human calculators. It was only later that the computer's more general capabilities as a symbol manipulator and a tool for automation were exploited.

Today computation is still the primary application of computers. We use them to compute wages, Social Security payments, taxes, inventories, invoices, bank balances, bridge cable thicknesses, aircraft wing shapes, spacecraft orbits, and practically everything else of numerical significance. And, just as in the days of ENIAC (Chapter 1, Figure 1), the newest, largest, fastest computers are still intended to solve computationally difficult, as opposed to logically difficult, problems. Today we refer to such machines as "supercomputers" or, somewhat more fondly, as "number crunchers," and we use them for such computationally awesome tasks as predicting the weather, predicting the effects of nuclear weapons, and designing experiments that may lead to the generation of electricity by nuclear fusion.

We really don't get very far on these problems by ourselves; at best, we have been able to solve them approximately, and even then only for severely scaled-down versions. Individually, the calculations involved are not difficult; we know perfectly well how to do them ourselves, but we just can't do them fast enough. Fortunately, we also know how to program computers to do the calculations, and computers *can* do them fast enough. In a minute or two a typical computer can do as many calculations as you or I can do in a lifetime. Supercomputers do them in a few seconds.

As information processors, however, humans are hardly helpless. Our brains may not be good computers, but they can solve enormously complicated information-processing problems. And just as computers routinely compute solutions to problems that remain beyond the reach of today's brains, we routinely solve problems that remain beyond the reach of today's computers—problems in natural-language understanding, pattern recognition, abstract reasoning, and the like. At best, we have been able to program computers to solve these problems approximately, and even then only for severely scaled-down versions.

Understanding Natural Language

In many ways this is the best example of a task that's easy for us but hard for computers. The ability to process natural languages with computers at similar speeds and with similar levels of understanding as humans is a long-sought-after goal. No one is certain whether this goal is feasible, but it's clear that the goal is currently out of our reach. The reasons have less to do with the power of available computers than they do with the current state of knowledge about natural languages and how people process them.

Consider syntax. Most of us learned some formal English grammar in school. Perhaps you had to spend time, as I did, diagramming English sentences in order to display their syntactical structure. I remember finding all sorts of sentences that I couldn't diagram. They didn't quite fit into the scheme of our grammar book, and I remember having inconclusive and frustrating discussions about them with my teacher. I wish he had just told me that the grammar book was incomplete and probably incorrect. In fact, linguists haven't been able to describe English syntax fully, and they don't really know how. There exist theories of English syntax, and there are large subsets of English for which there are correct and complete syntaxes. But although we rarely have much trouble with English syntax when we speak or write, we don't have a formal description of all the rules we are somehow able to follow. Because we don't understand the rules ourselves, we can't program computers to follow them.

The problems with semantics are much worse. For one thing, it's easy to underestimate how rich are the semantics of even a simple sentence. In Chapter 6 I mentioned some typical implications of the sentence "John drinks coffee," namely that a person named John sometimes pours a liquid into his mouth and swallows; that the liquid is made from the roasted and ground seeds of a particular plant that is grown primarily in South and Central America; that the liquid is probably hot and probably dark; and that John probably ingests caffeine when he drinks the liquid. But much more meaning is implicit in the sentence, and I'm sure that you can draw additional conclusions yourself—for example, conclusions about John's age, the relative probabilities of John's nationality or the regions in which he might live, and the type of container from which he drinks the coffee. You probably even noted that my name is John and considered the possibility that I was referring to myself.

Furthermore, in addition to their large size, the semantics of natural languages are laden with subtleties. Some of the conclusions I just mentioned involved subtle reasoning, but the subtleties involved were relatively simple. Here's another example: The sentence

"John finished the book."

is syntactically correct and semantically meaningful. The sentence

"The book finished John."

is also syntactically correct. On one level it's semantically meaningless, just like "Coffee drinks John," but on another, ironic level, it's not only meaningful but perhaps accurate. Another example of semantic subtlety is provided by a famous anecdote about a computer program that was designed to translate English automatically into Russian. When given the English sentence

The spirit is willing but the flesh is weak,

the program is said to have responded with the Russian equivalent of

The vodka is good but the meat is rotten.

Although this anecdote is probably apocryphal, it illustrates real complexities.

In general, the automatic analysis of natural language by computers is an extremely difficult problem, not just because it's complicated, but also because our own understanding of the numerous subtleties is limited. Although good work has resulted in modest progress, success for arbitrary discourse has not been achieved. It's one of those problems that looks much easier to solve than it is. For the past twenty to thirty years, experts have continued to assert that success was imminent while they continued to demonstrate otherwise.

Dealing with Ambiguity

I travel from Washington to New York several times a year. I prefer the train, but when I'm short of money or in need of punishment, I drive up Interstate 95 instead. Recently, a sign on the Delaware Memorial Bridge caught my eye. The sign—one of those signs with an army of regimented lightbulbs that can spell out any dozen letters or so—announced:

WIND CONDITIONS.

I absorbed its meaning quickly and drove on. Several minutes later I realized that the sign was ambiguous. CONDITIONS was meant as a noun, but it could be interpreted as a verb. What if I had seen those words on a billboard, together with a picture of a gorgeous head of luxurious hair? Is WIND a product of nature or a product of the petrochemical industry? Or is it a verb?

It's easy enough to explain why the meaning of the bridge sign is obvious to any driver, just as it's easy to explain why the different meaning of the hypothetical billboard would also be obvious. But it's hard to explain in general how people deal so successfully with linguistic ambiguity. Examples are helpful, but they don't define the general process. One can say that we resolve the ambiguities by a reasoning process that takes into account their contexts as well as all sorts of possibly relevant information from our past experiences. But this doesn't describe the process sufficiently to teach it to another person or to form the basis of a computer program.

Understanding natural language constantly requires the resolution of ambiguities. For example, the word "finished" has two common meanings: completed and killed. A few paragraphs ago, you resolved the ambiguity in the sentence

"John finished the book."

by noting that books are inanimate objects, not subject to killing. (Yet had the sentence appeared in a description of a parental mob descending angrily on a high school library's collection of adult books . . .) Similarly, you resolved the ambiguity in the sentence

"The book finished John."

by noting that, since books can't complete people, the more ironic meaning must apply. In choosing the ironic interpretation, you probably rejected a literal interpretation involving death. (Yet had that parental mob pelted John the librarian with those offensive books . . .)

Ambiguities in our daily experience are by no means limited to natural-language expressions. Ambiguity exists whenever information can be interpreted in more than one way. The information could be data from a scientific experiment, evidence about a crime, or results from a set of medical tests. Routinely, and often unconsciously, we consider the most likely interpretations and we select one of them. We do it so often and so naturally that we take the ability for granted. It's easy to expect computers to act likewise. They don't.

Proving Mathematical Theorems

The correctness of a computer program is evaluated by deriving the meaning of the program, as implied by the program's written text and the programming-language semantics, and comparing the results with the program's specification. Correctness can be expressed as a mathematical theorem stating that the program's meaning is consistent with the specification, and a mathematical proof of correctness can be carried out by proving such a theorem. Unfortunately, the proof involves formulas that quickly become unmanageably large and complicated for anything but the most trivial programs. Even though the theorem can be proved by performing a systematic series of transformations, substitutions, and rearrangements, to do it is tedious, time comsuming, and error-prone.

Such large formulas and such systematic procedures for analyzing them practically beg for computers to be used in proving theorems automatically. Indeed, interest in formal proofs of program correctness has stimulated much research into automatic theorem-proving. The results so far have been mixed. Program-proving systems have been developed, but their capabilities are limited. In almost all cases these systems can not be used to prove the correctness of realistic, useful software.

One reason is the limitations of programming-language semantics—the mathematical statements that define the meaning of each language statement. If a program includes language statements for which there are no formal semantics, then the program's meaning cannot be expressed in precise mathematical terms, and the meaning cannot be analyzed by an automatic program verification system. Unfortunately, most practical programming languages include such statements. For example, it's relatively easy to define formal semantics for numerical calculations involving integers. But for calculations involving non-integers, it's extremely difficult to define formal semantics because many such numbers are represented in computers only approximately.

Another reason for the limited capabilities of automatic theorem-provers is the complexity of the formulas that they have to deal with. I mentioned that computers came into play because the complexity of these formulas was too much for people; it turns out that computers are overwhelmed too! This happens because the logic in computer programs typically involves the consideration of many different combinations of conditions, which in turn causes the verification formulas to grow rapidly. This phenomenon is known appropriately as **combinatorial explosion.** Combinatorial explosion tends not to be a problem when people argue about mathematical theorems and program correctness, because people use methods that restrict the alternatives that must be considered at each stage. Our minds being limited, we really have no other choice.

Why not just write programs that prove theorems in the way that people do it? Unfortunately, we don't understand enough about how people do it. This may seem surprising given the reputation of mathematics as precise, systematic, and well-understood. But while these attributes apply to the results of mathematics, and even to the manner in which we present those results, they do not apply to the process by which we obtain them. We manage to prove mathematical theorems and argue convincingly about program correctness in all sorts of ways that have eluded automation.

Recognizing Patterns

Your friend is running in a marathon and you've volunteered to cheer and pass an orange. You stand on the curb, watching the running hordes. You see your friend, yell appropriately, and pass the orange. No problem. We recognize people easily, and not just when they face us motionless and alone against a bland background. No one has been able to program a computer to perform nearly as well.

Computer programs do process pictures in sophisticated and useful ways. Computers are used for deblurring and other forms of visual enhancement. Computer processing of TV images can assist industrial robots in selecting parts and positioning them correctly for installation. And computer programs can extract information from medical X-rays and weather satellite photographs. But computers haven't come close to matching our own ability to extract information from visual images. The problem is not so much a lack of processing power as it is a lack of knowledge. We simply do not understand well how we do it ourselves.

Consider the example of recognizing human faces—even in the restricted case of a single face against a bland background. A typical computer-based approach might proceed by measuring various obvious visual features numerically and comparing the results with stored sets of measurements. Chances are that such an approach would fail if someone switched from eyeglasses to contacts, shaved a mustache or beard, or changed a hairstyle, because the program would detect the changes and be confused by them. People can be fooled by a thorough disguise, of course, but most people aren't confused by changes such as these. On the contrary, people often don't even notice such changes or are aware of them only vaguely— "Hi, John . . . something's different about you, but I'm not sure what"—which shows that recognition is based on subtle features. More evidence for the subtleties and for the mysterious way in which we learn to detect them comes from the semi-serious statement "All _____ people look alike" (fill in the blank), which you're as likely to hear in Shaker Heights as you are in Harlem or Beijing.

We're just as good at recognizing patterns in sound. We recognize the difference between Preludes by Chopin and Bach, we recognize the difference between sad voices and happy voices, and we recognize what our friends say to us. Today you can buy special-purpose computers that will recognize around 100 different isolated words or phrases. If you speak one of the words or phrases, the computer can report, quite reliably, which one you spoke. Recognizing unrestricted continuous speech, however, is another matter. It might seem that humans recognize and understand speech by first transcribing the sounds mentally and then applying our knowledge of syntax and semantics. Indeed, early attempts to have computers transcribe speech directly into printed text were based on this assumption. These attempts approached the problem as you or I do if we're asked to write down the equivalent of a series of spoken nonsense sounds. The attempts failed miserably, and it became clear that knowledge of language syntax and semantics must come into play almost immediately. How else could people deal so easily with such packed examples as this:

I'll march down the aisle and marry Mary if Mary is merry.

More recent computer-based approaches to continuous speech recognition do exploit syntax and semantics. The results have been somewhat successful for severely restricted subsets of English, and it seems clear that they will lead to useful speech recognizers. But matching human performance remains well out of the computer's reach.

Our pattern recognition ability is a general one. It's not limited to sight and sound, and it can involve many senses at once. For example, many people can diagnose a car's problem by how it feels and sounds on the road. Others can identify a ripe cantaloupe by looking at it, shaking, pressing, and smelling it. Some doctors are better diagnosticians than others. Abilities like these are mysterious, poorly understood, and hard to imitate with computers.

Integrating Knowledge

We have programmed computers to do many things that we do ourselves. Computers can fly airplanes, analyze pictures, recognize spoken words, converse in subsets of natural languages, play chess, reason logically about specific subjects, compute the results of mathematical formulas, predict the weather, and attempt to cook beef Wellington. There are specific performance differences: people do some of these things better or faster than computers, and *vice versa*, but there's a more general difference. People are more than the sum of their particular abilities. As walking, talking, problem-solving beings, you and I integrate all of our specific abilities into something greater. But we don't know how to write computer programs that perform similarly.

Although the problems of knowledge integration are formidable and unsolved, many people assume that it would be a straightforward process to integrate the isolated successes of individual computer programs into a general-purpose robot system. Not so. A major unsolved problem is how to determine what's relevant. When an intelligent system encounters a problem, how can it determine enough about the nature of the problem to apply whatever relevant rules and algorithms are at its disposal?

A related problem is that of learning from experience. Although we know how to write computer programs that learn from their experience with a particular activity, we don't know how to write programs that exploit the resulting knowledge when they turn to a new activity. In this respect, people are much more capable than computers. When children learn how to stack blocks, they take advantage of whatever general problem-solving abilities and physical intuition they have acquired so far. Moreover, the process of learning how to stack blocks itself adds to the children's problem-solving abilities and physical intuition, which in turn can be applied to other activities.

Computers don't look like brains, they don't work like brains, and in many respects they don't perform like brains. The computer is the most important machine of our time, but its image as an electronic brain is misleading. In the words of Seymour Cray, the man who designed most of the supercomputers in use today:

> I don't know what all this supercomputer talk is all about. They certainly aren't supercomputers; they are kind of simple, dumb things.

Of course, computers do have their own advantages. Indeed, it's no exaggeration to say that computers have some superhuman capabilities, but so do other machines that we build. We can't fly as far as an airplane, lift as much as a derrick, or count as fast as a computer. It's true that today's brains are poor computers. But so are today's computers poor brains.

Computer or Knowledge Engine?

Computers may not be electronic brains, but they are electronic marvels. Tomorrow's computers will be even more marvelous; they'll store more information, perform faster, and perform better. Everyone agrees that these quantitative improvements will occur, but some also predict qualitative changes—they believe that tomorrow's computers will behave like today's people.

The Artificial Intelligentsia

This belief is central to "artificial intelligence" (AI), a branch of computer science that has an undeserved but large sway with the press and the public at large. Typical AI programs play chess, build cars, converse in natural languages, or diagnose diseases, activities that we tend to associate more with people than with machines. AI is the source of the most dramatic predictions about what computers will be doing in ten years. It is, for example, the main focus of Japan's Fifth Generation Computer Project, a billion-dollar attempt at creating technology to dominate the information age. If computers are the wave of the future, many see AI as riding the crest.

What, then, is AI? Unfortunately, that's hard to say, largely because the word "intelligence" is as problematical as the word "thinking." But it's fair to call **artificial intelligence** (AI) a field that concerns the use of machines to emulate human intellectual activities. This doesn't really pin it down, however, because the definition often doesn't determine whether or not a particular machine exhibits artificial intelligence.

One of my favorite examples of "AI or not AI?" is the autopilot. The pilot in me is impressed by any autopilot, even the relatively simple ones I've used myself. But when I read that an autopilot can make a Boeing 747 take off, fly to its destination, and land, more or less regardless of the weather and without the pilot having to do anything other than key in some numbers before takeoff and push an occasional button while enroute, I'm not just impressed, I'm amazed. If humanlike performance were the criterion, autopilots would qualify as AI devices. If you doubt this, you've never tried to land a plane in bad weather.

So why isn't an autopilot an example of AI? The simple, if cynical, answer is that autopilots preceded AI. They were first invented and are today produced by people from the more traditional fields of electrical engineering, aerodynamics, and control theory—people who don't regard themselves as being involved in AI. From their point of view, the autopilot is a device for solving some relevant equations from aerodynamics and control theory. History aside, it's clear that the autopiloting problem is treated as one of computation rather than one of thought. And for many in the AI field, that's enough to disqualify autopilots. To qualify, an autopilot would somehow have to reason about navigation problems, not just solve them.

This attitude is characteristic of the "cognitive modeling" school of AI. People in this school claim to use methods of representing and analyzing data that model human cognitive activity. Their programs attempt not only to produce results that are similar to those produced by people, but to produce them in the same way that people do. Now this is an interesting, even worthy goal; it offers a plausible route to better performance. But experiments to determine human thought processes are extremely difficult to do well, and few in AI spend much time trying. In most cases, the statement that a particular program reasons in the same way that people do is an assertion based on conjecture rather than evidence.

There is another school of AI, one more oriented to traditional engineering. For people in this school, what counts is being able to solve intellectually challenging tasks with computers, regardless of the methods used. Autopilots really do qualify for this school, even if they aren't admitted. Indeed, the mechanical control of robots is often accomplished with the same principles and methods as those used in autopilots. Yet robots are considered to be part of AI technology, while autopilots, even in pilotless aircraft, are not. The reasons have more to do with style than substance; AI sees itself as finding new solutions, and not as refining old ones. Robots are new and glamorous; autopilots are routine. In this respect autopilots are hardly unique; we take for granted all sorts of computer applications that must have seemed wondrous when they were first developed. To an extent, artificial intelligence is just another name for the routine methods of tomorrow.

In the end it all boils down to performance. Given a computer that's touted as having a particular human capability—whether it be thinking, making jokes, loving, or flying planes—the only reasonable test is to compare the computer's behavior with that of a person and see if you can tell which one is which. For example, you might sit down in front of a computer terminal and converse, by exchanging messages, with something in the next room. Your messages may contain whatever you want—commands, statements, questions, answers. Your task is to decide what's in the next room. Is it a computer, generating messages with an AI program? Or is it a person, typing messages on another terminal? This procedure is called **Turing's Test**, named in honor of British mathematician Alan M. Turing, who suggested it over thirty years ago in his classic paper "Computing Machinery and Intelligence."

Turing's Test bypasses questions about the computer's ability to think and goes to the heart of the matter: what computers can accomplish. It's clear that today's computers are a long way from passing Turing's test. But computers already accomplish a great deal, and it's conceivable that tomorrow's computers will confirm some of the dramatic predictions about AI.

The goal of AI is clear: to enable computers to perform well on tasks that today are the exclusive domain of the human intellect. Within the AI community, there's little doubt that this goal will be met. But the goal is incredibly difficult, and—as was true of alchemy in the Middle Ages—optimism is insufficient for success. Will AI provide the richest payoff of computer technology, or is it high-tech alchemy? Will AI transform electronic servants into savants? To judge requires more information about AI. How does AI work? What distinguishes AI programs from other programs? And what major obstacles stand firmly between the artificial intelligentsia and its goals?

Symbol Manipulation and AI

The key characteristic of AI programming is its dependence on symbol manipulation rather than on numerical calculation. Word processing is a relatively simple example of symbol manipulation. In word-processing software, bit patterns in memory are interpreted as text characters rather than as numbers. But there's no need to stop at text characters. You can choose to interpret bit patterns as symbols for objects (bread, butter, jam, jar, knife, sandwich), for actions (open, cut, dip, spread), for relationships (on top of, underneath, inside of), for attributes (rye, soft, stale, sharp), and for anything else you care to suggest. Moreover, you can write programs that manipulate the bit patterns in ways that are consistent with their symbolic meanings. You could write a program to determine whether there are enough ingredients around to make a sandwich. And you could write another program that would, symbolically at least, make a sandwich. Generally speaking, such programs are symbol manipulation programs.

In symbol manipulation programs, it's convenient to represent information as lists of symbols like

(bread butter filling butter bread).

Additional generality is achieved by permitting symbols to stand for other lists—for example, filling could stand for the list

(cheese ham lettuce).

In the lists above, each symbol has at most two neighbors; for example, ham has the neighbors cheese and lettuce. Even more generality results from structures in which symbols can have more than two neighbors. In these general structures, called **graphs** or **networks**, there can be many connections between symbols and more than one kind of connection. The different kinds of connections are themselves represented as symbols that indicate different relationships among the connected symbols (e.g., on top of, in between, tastes like).

Elaborate symbolic structures can be used to represent any object or concept: a sandwich, a chess board, a mathematical theorem, a family tree comprising seven generations, an automobile, the layout of a room full of furniture, a map. They can also be used to represent the actions, rules, and strategies involved in some procedure: making a sandwich, playing chess, navigating through a room.

For years the term "data base" has been used in the context of computer systems to refer to any large file of information. The symbolic structures I just described are also data bases, but the AI community refers to them as **knowledge bases**. It appears that the distinction arose to emphasize the symbolic over the numeric and the inferential over the factual. But traditional data bases do contain symbols and they are used for inference. Moreover, in my *Concise Oxford Dictionary* I find

dāt´ um, n. (pl, *-ta*) Thing known or granted, assumption or premise from which inferences may be drawn.

The distinction between "knowledge base" and "data base" is more psychological than real: it's a bit like saying user-friendly instead of idiot-proof, or Palmetto bug instead of cockroach.

By creating, extending, searching, and otherwise manipulating symbolic information structures, one can solve specific information-processing problems. A variety of so-called AI programming techniques have been developed to help do this. For example, **goal-directed searching** techniques facilitate the discovery of procedures that meet a goal stated in symbolic terms, e.g., "win the game" or "navigate to New York". An important attribute of these techniques is their ability to retreat from unpromising search paths without starting entirely over, a procedure known generally as **backtracking.** Another important attribute is the ability to exploit **heuristics,** sort of algorithmic rules-of-thumb that guide the search for a goal. People use heuristics all the time: "When lost, stick to the main road." "The way to a man's heart is through his stomach." We know the heuristics aren't guaranteed to work, but we believe that they usually help.

AI programming methods are facilitated by special programming languages. Although you can use them to compute numerical quantities, these languages are oriented more to symbol manipulation than to numerical computation. Because one of their fundamental purposes is to facilitate defining and manipulating lists of symbols, they are referred to generically as **list-processing languages.** The most famous and influential of these is **Lisp,** short for **list** processing, a language that was developed by the American computer scientist John McCarthy and others during the late 1950s and early 1960s. Another important example is **Prolog,** the programming language that the Japanese have chosen for their Fifth Generation Computer Project. One of the properties that distinguishes Lisp and related languages is that Lisp programs are themselves represented as lists of symbols. This makes it easy to write Lisp programs that create and run other Lisp programs, a powerful and intriguing capability that is used extensively in AI programming.

One of the most important classes of AI programs today are known as **expert systems**—programs that attempt to emulate the behavior of an expert in some particular field, say, internal medicine. Given a specific problem—say, diagnosing a disease from the results of medical tests—expert systems produce an answer by exploiting a large number of rules and heuristics. The rules and heuristics are programmed in after studying relevant literature and interviewing experts. Two famous examples of expert systems are SYNCHEM, a program that proposes procedures for synthesizing organic molecules, and DENDRAL, a program that proposes chemical structures based on data from an instrument known as a nuclear magnetic resonance spectroscope. A more recent example is XCON, a program used by the Digital Equipment Corporation to complete the configuration of computer systems—down to parts lists that include nuts, bolts, and cables—based on approximate descriptions of the systems.

To describe the process of designing, building, and testing expert systems and related AI software, the AI community has coined the term **knowledge engineering.** This pretentious term does have some advantages: it emphasizes that more than programming is involved; it's consistent with the term "knowledge base"; and it makes engineering sound more glamorous than usual. Indeed, knowledge engineering is widely touted as the most significant new result of computer technology, and it is the basis for a growing number of commercial ventures.

Trust in Expert Systems

As expert systems move from the research world, where people experiment with them, to the commercial world, where people will depend on them, it becomes important to consider their reliability. Unfortunately, while expert systems may be more glamorous than other software, they are not more reliable. On the contrary—like other large computer programs, expert systems suffer from the inadequacy of testing, the curse of flexibility, the existence of invisible interfaces, and other problems that helped to bring on the software crisis. In addition to such "classical" problems, however, expert systems have a related one of their own: The rules and heuristics that determine their behavior interact in complicated and unpredictable ways; changes that improve an expert system's performance on one problem can easily degrade its performance on another problem. As a result, managing complexity is even more of a problem in expert systems than it is in many other kinds of software.

We know that software reliability depends on the precise specification of required behavior. We know that reliable software is designed for ease of change. We know that software reliability depends on keeping software within our intellectual grasp. We know that testing is inadequate for proving software correctness. And we know that adequate proofs of correctness must consider the written text of the software. Unfortunately, these important lessons of the software crisis are, with few exceptions, disregarded during the design and construction of expert systems. One reason for this disregard is just the technological aloofness of those who believe themselves to be at the pinnacle of high-tech. Another reason arises from the way in which expert systems are designed and constructed. After reading relevant technical literature and interviewing one or more experts, the "knowledge engineer" formulates an initial set of rules and heuristics and then writes a computer program embodying them. This program serves as a prototype of the expert system. How is the final system derived from the prototype? To quote from a popular textbook on building expert systems:

> Refinement of the prototype normally involves recycling through the implementation and testing stages in order to tune or adjust the rules and their control structures until the expected behavior is obtained.

In short, reliability is supposed to result from testing. Testing is inadequate in general, but the situation is even worse in the case of expert systems, as the correct behavior in any particular case is often a matter of opinion. Moreover, because "expected behavior" is not specified completely and precisely, there's no way—*even in principle*—to prove the correctness of these expert systems.

I'm not opposed to the goals of those who build expert systems today—only to their methods. Programs that emulate human intellectual activities are both conceivable and worth developing. But they should be subjected to the same discipline of specification and construction that the software crisis has demonstrated to be imperative. How to write adequate specifications for such advanced computer systems is not clear, but there are hints in the same place that provides the most fantastic predictions of artificial intelligence: the science-fiction literature. There, a characteristic embodiment of AI is the mobile robot that can communicate in a natural language and carry out a wide range of tasks. To deal with one aspect of robot reliability, Isaac Asimov introduced the Three Laws of Robotics:

  I. A robot may not injure a human being or, through inaction, allow a human being to come to harm.
  II. A robot must obey the orders given it by human beings, except where such orders would conflict with the first law.
  III. A robot must protect its own existence, as long as such protection does not conflict with the first or second law.

The Laws have become somewhat of a fixture in the science-fiction literature. Where they apply, people can count on them.

Why people can count on them has not really been addressed, but they are supposed to be guaranteed as part of the process by which robots are designed and developed. Although the precise definition of such terms as "harm" is problematical, the Laws can be viewed as an informal specification. With a little imagination (comes with the genre), one can think of stating some similar requirements in mathematical terms, thereby providing the basis for a formal proof of correctness. This is not as far-fetched as it might sound. There are precise mathematical formulations of such informal concepts as "consistent inference" and "logical deduction." It's conceivable, even today, to state formally the general requirement that expert systems should draw only logical conclusions from the data presented to them. A hundred years from now, people may laugh at accounts of circa 1985 "expert systems" that were built by programmers who didn't even think about precise specifications and proofs of correctness.

The situation today reminds me of Knuth's description of programming as it existed twenty-five years ago: "We just fiddled with a program until we 'knew' it worked" (for the full quote see pages 205 and 206). Building expert systems by trial and error is a technological giant step, backwards; to ignore the lessons of the software crisis is scientifically shallow and socially irresponsible. Without software engineering, the knowledge engine will backfire.

Crunching Numbers and Bashing Symbols Isn't Enough

A ten-million-dollar supercomputer that can predict next week's weather is pretty impressive. It would be even more impressive if it could predict the next great California earthquake, but it can't, at least not yet. Why the difference in predictive capabilities? Simple. It doesn't have to do with the computer, it has to do with us. We know how to predict weather, but we don't know how to predict earthquakes; we don't understand enough about the causes of earthquakes.

In general, our use of computers to solve a given problem requires two things: sufficient understanding to solve the problem, and sufficient electronics to proceed with the solution. Of the two, understanding is clearly more important. Without supercomputers it's probable that we couldn't obtain good, long-range weather predictions. But it's certain that we couldn't obtain them without the substantial understanding of atmospheric physics that makes it possible to write the computer programs.

In 1979 I was asked to write a short position paper on the relevance of AI technology to the Navy. The problems and opportunities were obvious: The Navy spends a huge amount on military personnel, operations, and maintenance; in 1979, it was already $22 billion. The Navy's dependence on computers is growing rapidly. Not only are computers required by the speed and technical sophistication of modern warfare, they also have the potential for saving personnel costs—in contrast to projections about Naval personnel, computers will continue to decrease in cost and increase in capability. Of all the branches of computer science and engineering, AI stands out as promising the most. Perhaps a large investment in AI would pay off handsomely.

I was ambivalent. Many useful and popular methods for symbol manipulation were first developed in AI projects, and AI projects continue to pioneer additional methods. I was impressed by the newer methods for defining and manipulating elaborate networks of symbols. I was also impressed by some of the programs that had been built with them. But many other AI projects didn't impress me at all. These projects—and many similar ones exist today— were experiments without stated hypotheses, without objective measurements of success or failure, and without the intellectual discipline that is so important for successful scientific investigations. When the projects are over, it's unclear what has been learned. Rarely are there general conclusions that can be passed on to others. And when success is claimed, rarely can the results be applied to practical problems; usually the results were obtained for oversimplified problems—called "toy" problems by all involved—and they no longer work when the problem is scaled up to realistic proportions.

As I prepared to write the position paper, I was impressed by a conversation with the American computer scientist Alan J. Perlis. A distinguished figure in the history of symbol manipulation, Perlis is today involved in several AI projects. He said something to me that stuck:

> Good work in AI concerns the automation of things we know how to do, not the automation of things we would like to know how to do.

I realized why I had been so ambivalent.

Computers can help us to reap the benefits of understanding, and they can help us to achieve understanding, but they can't replace it. The AI projects I disliked were trying to produce programs that would appear to read like people, see like people, reason like people, and so on. In contrast to the projects that produced SYNCHEM, DENDRAL, and XCON—computer programs that automate relatively well-understood procedures—these other projects were trying to automate human capabilities without understanding them, and their outcomes were never much in doubt. In striving to leap so far, they didn't just fall short; they made no progress at all.

Does this mean that no one should ever try to make the attempt? Of course not. Outstanding individuals occasionally are capable of great intellectual leaps, and we may yet see such a leap here. But these are exceptions; the rest of us must either approach such problems with humility or be prepared to waste our time.

Automation vs. Augmentation

For many people, the issue isn't whether computers can replace understanding—it's whether computers can replace people.

Concern about this issue doesn't require faith in the technological promises of the artificial intelligentsia—one need look no further than today's factories, offices, and schools, where computers are doing more and more things that people used to do. It's a trend that will continue, and people wonder where it will lead. Pessimistic predictions focus on the computer's capability for automation. I think it's the wrong focus. The computer's most significant capability isn't automation, it's augmentation.

Personal computers are a good example. Their popularity is largely the result of three types of software: games, word processing, and spreadsheets. There is, of course, a degree of automation in all of these. But the automation isn't there to replace us. We interact with the computer across a user-interface, exploiting the automation to augment our capabilities. Computer games challenge me. A word processor helped me to write this book faster and better. And when I bought a new car recently, I used a spreadsheet to help me understand the financial tradeoffs. I watched the monthly payment change as I changed the price of the car, the down payment, the interest rate, and the number of payments. The experience helped me make a more intelligent purchase. The computer doesn't replace me, it enhances me.

Whenever a computer automates one of our jobs, we are free to take on another one. And when we do, the computer itself is often our most useful tool. In science, art, education, and business, computers free us from the mundane and assist us in new achievements. The computer is a tool for automation, but it's also a tool for exploration, illumination, and comprehension. It is a tool for intellectual leverage. The computer is important not because it allows us to do less, but because it helps us to do more.

Notes

Chapter 1: Intimidation and Anxiety

page

5   If you are interested in reading more about the history of computers, I suggest *The Computer from Pascal to Von Neumann*, by Herman H. Goldstine (Princeton, Princeton University Press, 1972).

7   For a readable account of some issues related to the process of belief, including experiments that show the strength of our urge to believe in psychic phenomena and magic, see Douglas R. Hofstadter's "Metamagical Themas" column in the February 1982 edition of *Scientific American*.

8   Richard P. Feynman, "There's Plenty of Room at the Bottom" (transcript of a talk given by Dr. Feynman on December 29, 1959, at the annual meeting of the American Physical Society).

15   The angstrom is a tiny unit of measurement named after the Swedish physicist A. J. Ångström; in one centimeter there are one hundred million angstroms.

15   In our daily experiences we encounter huge objects and tiny objects, but their range of sizes is miniscule when compared to the range of sizes extending from atoms to the universe. For a fascinating and intuition-building photographic account of the range of sizes in the universe, see the book *Powers of Ten*, by Philip and Phyllis Morrison (W. H. Freeman & Company, San Francisco, 1982).

15   Douglas R. Hofstadter discusses the T4's behavior in the context of computer science, mathematics, music, art, and whimsy in his book *Gödel, Escher, Bach: An Eternal Golden Braid*, New York, Basic Books, 1979.

25   The Robert S. McNamara quote is from *The Essence of Security* (1968).

page

29 Captain Hopper is quoted in the *Encyclopaedia of Computer Science and Engineering*, A. Ralston and E. D. Reilly, Jr., editors, New York, Van Nostrand Reinhold, 1983, p. 685.

30 Statistics on the role of women in computer science and engineering were obtained from *Professional Women and Minorities—A Manpower Data Resource Service*, B. Better and E. Babco (editors), Scientific Manpower Commission, Washington, D.C., 1983. The 1978 salary survey was summarized in *Women and Minorities in Science and Engineering*, National Science Foundation, NSF 82-302, Washington, D.C., 1982.

32 The origin of the Tallulah Bankhead quote is a bit mysterious. I found it in Bohle's *The Home Book of American Quotations*, which claims that Bankhead made the remark to Alexander Woollcott during a March 24, 1919, performance of Maeterlinck's *The Burgomaster of Stilemonde*, and that the remark was quoted in a *New York Times* review that was published the next day. I looked in *The New York Times* of March 25, 1919, and found the review of Maeterlinck's *The Burgomaster of Belgium* (Stilemonde is a city featured in the play), but no mention of either Woollcott or Bankhead! Perhaps there was more than one edition of the *Times*.

Chapter 2: Coming to Terms with Computer Jargon

page

45 W. Strunk, Jr., and E. B. White, *The Elements of Style*, New York, The Macmillan Company, 1972.

45 E. Gowers, *The Complete Plain Words*, Penguin Books, 1962.

45 Gowers, p. 20.

46 Although I have permission to quote this material, I've chosen not to identify it because I don't want criticism of a particular system to detract from the general point I'm making. Also, since I didn't study the system thoroughly, the example might be misleading about the system's overall quality (I doubt it). Elsewhere in the book I use other derogatory examples from popular personal computer systems, and for the same reasons I change their names or ignore their identities. An exception to this practice is my harsh treatment, in later chapters, of the Apple IIe computer combined with the Apple Writer II word-processing program. In this case, I used them both intensively for a few weeks, and I believe my criticisms to be fair. Also, for all the criticism I heap on the Apple IIe/Applewriter II combination, I heap as much praise on Apple's Lisa and Macintosh. Seems a fair trade.

49 Figure 8 is adapted from material used with the permission of Perfect Software, Inc., Berkeley, CA.

49 It's likely that Figure 8 was preceded by a formal specification written in a notation called a "metalanguage." The problems with Figure 8 could all arise from translating such a specification into English. Here, presented without explanation, is a metalanguage specification that could have led to Figure 8; unlike Figure 8, it is complete and unambiguous:

number = real-number | exp-real-number
real-number = [" – "] (digit {digit} ["."] | {digit} "." digit {digit})
exp-real-number = real-number ("E" | "e") (" + " | " – ") (digit | digit digit)
digit = "0" | "1" | "2" | "3" | "4" | "5" | "6" | "7" | "8" | "9"

The metalanguage is one that's used in a standard definition of the programming language Pascal. (*American National Standard Pascal Computer Programming Language*, ANSI/IEEE770x3.97-1983, Institute of Electrical and Electronic Engineers, New York, 1983) The metalanguage symbols are described in the following table:

Metalanguage Symbols

Metasymbol	Meaning
=	shall be defined to be
|	alternatively
[x]	0 or 1 instance of x
{x}	0 or more instances of x
(x | y)	grouping: either of x or y
"xyz"	the terminal symbol xyz

52 There's a long-standing tradition of explaining computers by means of anthropomorphic analogies. Indeed, when the modern stored program computer—also called the Von Neumann computer in honor of the American mathematician John Von Neumann—was described by Von Neumann and others in 1946, they used such anthropomorphic terms as "memory organ," "control organ," and "input and output organ."

56 Many people read that a large programming project required 200 programmer-years of effort and they conclude that programming is a highly social activity. In fact, the most successful methods for building such large programs are those that allow programmers to work in relative isolation. I talk more about this in Chapter 10.

57 A collection of computer slang has been published by Guy L. Steele, Jr., and numerous colleagues as *The Hacker's Dictionary—A Guide to*

page

the World of Computer Wizards (New York, Harper and Row, 1983). The entry for "hacker" includes seven definitions.

58 Grace M. Hopper, "The First Bug," *Annals of the History of Computing*, Vol. 3, No. 3, July 1981, pp. 285–86.

Chapter 4: User-Interfaces, Friendly and Unfriendly

page
83 Letter from Peter Buhr, University of Manitoba, Canada, published in the *Communications of the ACM*, Vol. 26, July 1983, pp. 463–64.

84 R. L. Mack, C. H. Lewis, and J. M. Carroll, "Learning to use word processors: problems and prospects," *ACM Transactions on Office Information Systems*, Vol. 1, July 1983, pp. 254–71.

Chapter 5: Facing the User-Interface

page
94 The figure of 260,000 species of vegetables actually refers only to tracheophytes (vascular plants).

95 George A. Miller, "The magical number seven, plus or minus two: some limits on our capacity for processing information," *The Psychological Review*, Vol. 63, No. 2, March 1956, pp. 81–97.

95 Alan Kay, whose work on user-interfaces strongly influenced the Star, Lisa, and Macintosh designs, was quoted in "Designing the Star user interface," by D. C. Smith et al. (*BYTE*, April 1983, pp. 242–82).

105 Charles Rubin, "Some people *should* be afraid of computers," *Personal Computing*, August 1983, pp. 55–57, etc.

Chapter 6: Conversing with Computers

page
109 Joseph Weizenbaum writes about his experience with ELIZA in his book *Computer Power and Human Reason* (W. H. Freeman & Company, San Francisco, 1976).

111 The phrase "do what I mean (DWIM)" has become a refrain of computer slang. It arose from a program called DWIM, written by Warren Teitelman, that attempted to correct faulty inputs.

113   I've heard different versions of the "Programmer's Lament" ditty. This one I adapted slightly from one printed in "Joy of Hacking" by F. Rose (*Science 82*, November 1982, pp. 59–66).

114   Unix was developed by Ken Thompson and Dennis Ritchie at Bell Laboratories in 1969–71 for the Digital Equipment Corporation (DEC) PDP-11 series of mini-computers. Since then, it has become popular on a variety of larger and smaller machines. Recently, unix and unix-like systems have begun to permeate the personal computer market as well.

122   In the notes to Chapter 2, I included a typical metalanguage definition of numbers. In fact, this was a formal definition of the syntax of numbers. Such metalanguage definitions are used for the syntax of many artificial languages, not just for numbers. An important advantage of such formal definitions is that they can be translated automatically into a computer program that will break expressions up into their constituent parts according to the defined syntax.

Chapter 7: The Sachertorte Algorithm

131   The programming language Pascal was designed in 1968 by the Swiss computer scientist Nicklaus Wirth. It has become a popular language for a wide range of programming, from large systems to small.

134   Irma S. Rombauer and Marion Rombauer Becker, *The Joy of Cooking*, Bobbs-Merrill, 1984.

139   The beef Wellington recipe is from *Mastering the Art of French Cooking*, Vol. II, by Julia Child and Simone Beck (New York, Alfred A. Knopf, 1973), pp. 181–185.

140   Amelia Bedelia is a literal-minded maid who was the main character in a series of children's books written by Peggy Parish. When Amelia Bedelia, for example, was told to "strip the sheets of the bed," she cheerfully tore them into narrow strips. And when she was told to "remove the spots from the dress," she did so, with scissors.

142   The bug in Algorithm AMT-4 is in the **if-then** statement just before statement 30. The opposite condition is what's needed, i.e.,

   if "egg yolks are left" **then goto** 20;

page

    instead of

    if "no more egg yolks" **then goto** 20;

Chapter 8: Myths of Correctness

page

162 The Ellen Goodman column was called "Whodunit? The Computer"; it appeared in *The Washington Post* of December 13, 1983.

170 William Kahan described "A Penny for Your Thoughts" in his paper "Mathematics written in sand—the hp-15c, Intel 808/, etc." (prepared for the Joint Statistical Meetings of the ASA-ENAR-WMAR-IMS-SSC held in Toronto, Canada, August 1983).

173 Dijkstra's remark about program testing appears in a variety of places. Its earliest generally-accessible publication was in his monograph "Notes on Structured Programming," which appeared in *Structured Programming*, by O.-J. Dahl, E. W. Dijkstra, and C. A. R. Hoare (London, Academic Press, 1972).

181 All of the Fred Brooks quotes are from his collection of essays *The Mythical Man Month* (Addison-Wesley, 1975). The original copyright date is 1972. This humble and entertaining collection of essays is still popular and still informative, which tells you something about progress in software engineering.

Chapter 9: Programming As a Literary Activity

page

185 Edsger Dijkstra's remarks about FORTRAN are from 1975. They became generally available in E. W. Dijkstra, *Selected Writings on Computing: A Personal Perspective*, New York, Springer-Verlag, 1982.

186 The Dijkstra quote about programming being a gigantic problem is from his 1972 Turing Award lecture "The Humble Programmer," which was published in *Communications of the ACM* (Vol. 15, No. 10, October 1972, pp. 859–66). Although published in a technical journal, this lecture can easily be understood by the uninitiated, and I recommend it if you're interested in reading more about the software crisis and about modern attitudes toward programming. The Turing Award is the most prestigious award in computer science. It was named for the

page

British mathematician Alan Turing, and has been given annually since 1966 by the Association for Computing Machinery (ACM). Other Turing Award recipients mentioned elsewhere in this book include John Backus, Robert Floyd, C. A. R. Hoare, John McCarthy, and Alan Perlis.

187   George Orwell, in *Collected Essays, Journalism, and Letters of George Orwell*, Sonia Orwell and Ian Angus, editors, New York, Harcourt Brace Jovanovich, 1968, Vol. 4, pp. 127–140.

188   The history of the **goto** controversy and the **goto**'s status in 1974 was summarized by Donald Knuth in "Structured Programming with **goto** Statements" (*Computing Surveys*, Vol. 6, No. 4, December 1974, pp. 261–301). Knuth gives many examples of how the judicious use of **goto**s can be beneficial for programs written in Pascal-like languages. Dijkstra's remark about not being "terribly dogmatical" is reported by Knuth.

190   The George Orwell quote is from "Politics and the English Language."

192   The Dijkstra quote about PL/I comes from his Turing Award lecture, cited above.

196   Brian W. Kernighan and P. J. Plauger, *The Elements of Programming Style*, Second Edition, New York, McGraw-Hill, 1978.

196   The mystery rule comes from Strunk and White. There is, of course, an analogous one in Kernighan and Plauger:
Write clearly—don't sacrifice clarity for "efficiency."

197   The book *Structured Programming*, by O.-J. Dahl, E. W. Dijkstra, and C. A. R. Hoare, was published by Academic Press (New York, 1972).

197   Knuth's remark about the book *Structured Programming* appeared in *Computing Surveys*, cited above.

197   The Hoare description of structured programming was quoted by Knuth in *Computing Surveys*, cited above.

Chapter 10: Programming As Mathematics, Programming As Architecture

page

199   If you're interested in the idea of programming as an art, you will enjoy Donald Knuth's 1974 Turing Award lecture "Computer Programming As an Art," which was published in *Communications of the ACM* (Vol. 17, No. 12, December 1974, pp. 667–73). Like Dijkstra's Turing Award lecture, Knuth's can easily be understood by the uninitiated.

page

205 Knuth's remarks about program proving are from his 1974 Turing Award lecture "Computer Programming As an Art," cited above.

208 Le Corbusier: *"La maison est une machine à habiter."*

209 The point about the telephone's effect on the economic viability of sky-scrapers is made in several places within *The Social Impact of the Telephone* (Ithiel de Sola Pool, ed., MIT Press, 1977).

220 The Winston Churchill quote is from *Time* magazine, September 12, 1960.

222 John Backus has described the history of FORTRAN in "The History of FORTRAN I, II, and III" (Richard L. Wexelblat, ed., *History of Programming Languages*, New York, Academic Press, 1981, pp. 25–74).

222 The Dijkstra quote is from his *Selected Writings on Computing*, cited earlier.

Chapter 11: Electronic Cretins

page

224 The B. F. Skinner quote is from *Contingencies of Reinforcement.*

224 *Webster's New Collegiate Dictionary*, G. and C. Merriam Company, Springfield, Mass., 1977.

225 President Kennedy is quoted in Simpson's *Contemporary Quotations.*

235 The Seymour Cray quote is from an interview published in *Datamation* (J. Johnson, "Four Expert Opinions," *Datamation*, December 1982, pp. 143–150).

236 Artificial Intelligentsia—a delicious expression, whether in reference to robots or their makers. The expression isn't mine. I've heard it in several places; Joseph Weizenbaum attributes it to Louis Fein.

238 Alan M. Turing, "Computing Machinery and Intelligence," *Mind*, Vol. LIX, No. 236 (1950).

242 The quote about testing expert systems is from *Building Expert Systems*, by Frederick Hayes-Roth, Donald A. Waterman, and Douglas B. Lenat (Addison-Wesley Publishing Co., Inc., Reading, Massachusetts, 1983).

243 *I, Robot*, by Isaac Asimov (Ballantine Books, 1947), is a stimulating (and thirty-five-year-old!) collection of stories about the Three Laws of Robotics. *The Robots of Dawn* (Doubleday, 1983) is a recent book by Asimov that features the Three Laws.

page
245 Nonscientists often assume that the "scientific method" is understood well and practiced uniformly by all scientists. Not true. Different scientists proceed in different ways, some less productively than others. My favorite discussion of the "right" way is contained in an article written by John R. Platt: "Strong Inference," *Science*, Vol. 146, No. 3642, October 16, 1964, pp. 347–53.

Index